CORROSION AND MARINE GROWTH ON OFFSHORE STRUCTURES

CORROSION AND MARINE GROWTH ON OFFSHORE STRUCTURES

J. R. LEWIS
Department of Chemistry
University of Aberdeen
and
A. D. MERCER
National Physical Laboratory
Teddington, Middlesex

Published for the
SOCIETY OF CHEMICAL INDUSTRY
London, by

ELLIS HORWOOD LIMITED
Publishers · Chichester

First published in 1984 by

ELLIS HORWOOD LIMITED
Market Cross House, Cooper Street, Chichester, West Sussex, PO19 1EB, England

The publisher's colophon is reproduced from James Gillison's drawing of the ancient Market Cross, Chichester.

Distributors:

Australia, New Zealand, South-east Asia:
Jacaranda-Wiley Ltd., Jacaranda Press,
JOHN WILEY & SONS INC.,
G.P.O. Box 859, Brisbane, Queensland 40001, Australia

Canada:
JOHN WILEY & SONS CANADA LIMITED
22 Worcester Road, Rexdale, Ontario, Canada.

Europe, Africa:
JOHN WILEY & SONS LIMITED
Baffins Lane, Chichester, West Sussex, England.

North and South America and the rest of the world:
Halsted Press: a division of
JOHN WILEY & SONS
605 Third Avenue, New York, N.Y. 10016, U.S.A.

British Library Cataloguing in Publication Data
Corrosion and marine growth on offshore structures
1. Office structures 2. Sea-water corrosion
I. Lewis, J.R. II. Mercer, A.D.
627'.98 TA462

Library of Congress Card No. 83-18637

ISBN 0-85312-564-3 (Ellis Horwood Limited)
ISBN 0-470-27506-5 (Halsted Press)

Typeset by Ellis Horwood Limited.
Printed in Great Britain by R.J. Acford, Chichester.

Table of Contents

CHAPTER 1

Marine growth and corrosion

1.1 INTRODUCTION

This volume contains the papers and discussion from a meeting on the subject of marine corrosion of offshore structures held at the Chemistry Department of the University of Aberdeen in September 1982. The meeting was organised jointly by the Aberdeen Section and the Materials Preservation Group of the Society of Chemical Industry and was the successor to a previous meeting on this subject held under the same auspices in 1979.†

There have been numerous conferences on the subject of marine corrosion and some of these have concentrated on aspects relating to offshore structures for oil and gas production. However, the intention in this conference was to broaden the approach by including substantial contributions on the nature of the marine growths that can develop on such structures and how these growths might affect the corrosion processes and the protective measures that are employed. To that extent it is believed that the material in this volume, particularly when taken in conjunction with the earlier publication, provides a valuable blend of data on biological and corrosion effects on steel structures in a marine environment.

The term 'marine growth' is interpreted widely to include all forms of biological activity, for example bacteria, barnacles, seaweed, etc. The whole range of these activities can potentially affect corrosion and protection processes, although, as the papers will show, the connections are not always proven.

This volume contains three chapters. The first, which consists of the introduction and five papers, is devoted to what may be described as the problem, that is, marine growth and corrosion. In this chapter the subject of marine fouling is discussed in all its various aspects. There are contributions on microfouling, microalgal fouling, sulphate-reducing bacteria and their effects on the environment of steel and concrete structures. It appears that early

† The volume based on the first conference is available from the Society of Chemical Industry: ISBN 090100 170 8, at a price of £16 (Society Members £8).

predictions of less marine growth in the northern North Sea than in the central North Sea have not been fulfilled since fouling is occurring to similar extents in the two regions. However, the types of growth are unpredictable and can differ significantly even on neighbouring platforms. These various forms of biological activity are considered in detail in separate papers in this chapter.

The influence of sulphate-reducing bacteria on corrosion processes has been known for many years but the quantitative assessments of their activity have not kept pace with the more frequent descriptive accounts of the phenomena associated with their action. One paper is devoted to a sulphate-reduction assay which, it is expected, will provide a rapid, accurate and direct measure of the critical biological activity that is the direct cause of anaerobic corrosion. In other papers the effects of these bacteria on corrosion and hydrogen sulphide production inside and outside offshore structures are considered with quantitative data supplied for distribution around a platform. Apart from bacterial effects, the microalgae can also affect the corrosion of steel since these produce copious mucilage which, on the one hand, may have beneficial effects in consolidating scales or corrosion products or, on the other, set up local micro-climates that will encourage bacterial activity, localised corrosion, etc.

With offshore developments now in the waters off the west of the British Isles it is important that studies of corrosion and fouling should be initiated in these areas. Such information is required, for example, in connection with the National Wave Energy Programme and preliminary results from studies in the western coastal (Hebridean) regions of Scotland are included in this chapter.

The second chapter considers inspection techniques, the practical problems of cathodic protection, the possible effects of this type of protection on metal properties, protection by painting and the associated cost implications, and the use of corrosion-resistant materials – including stainless steel – in offshore work.

It is emphasised that the main objective of underwater inspection is to locate, quantify and report deterioration. There is a need for divers to be directed by engineers with experience of corrosion and knowledge of the structure and its environment. One paper describes the difficulties that can be experienced in the installation, operation and maintenance of impressed current cathodic protection schemes. Successful protection of the structure was achieved despite the many problems that are graphically recounted.

The possibility of hydrogen pick-up by steels from the application of cathodic protection (CP) schemes is always of concern since adverse effects on material properties may result. It is now considered – from the available information – that normal levels of CP will not be significant in causing failures either by stress corrosion or corrosion fatigue even at welded joints. The author of the paper concerned with these aspects points out that more work is still needed on the effects of over-protection and of sulphate-reducing bacteria on hydrogen pick-up in CP conditions.

The difficulties to be expected in sea-water circulating systems on offshore structures are also described: examples are given of the care that is needed in the design of equipment to avoid the problems arising from the presence of crevices and bimetallic effects.

Significant corrosion problems also arise in the splash zones and topsides of platforms. In both areas good painting schemes can provide good protection, but the costs can be considerable, particularly when working in and above the splash zone, and the reasons for these are considered in some detail in one of the papers in this chapter. An alternative approach for the topside structure is to use stainless steels and examples of this application are given with interesting cost comparisons.

Although anti-fouling coatings have been used in protection of ships for many years less attention has perhaps been given to their use for static structures. New developments in this area are described with an account of new coatings which make 'adhesion' of growths more difficult.

The final chapter presents a summary of the various discussions that took place after the formal presentations of the papers. This material serves to amplify some of the points made by the authors in Chapters 1 and 2.

J. R. Lewis
Aberdeen Section, SCI;
Chemistry Department,
University of Aberdeen, UK

A. D. Mercer
Materials Preservation Group, SCI;
National Physical Laboratory,
Teddington, Middlesex, UK

1.2 PATTERNS OF MACROFOULING ON STEEL PLATFORMS IN THE CENTRAL AND NORTHERN NORTH SEA

G. N. R. FORTEATH, G. B. PICKEN and **R. RALPH**, Offshore Marine Studies Unit, Zoology Department, University of Aberdeen, Aberdeen, Scotland

ABSTRACT

The Offshore Marine Studies Unit has studied marine growth on steel installations placed in twelve oilfields in the central and northern North Sea. Samples, colour photographs and video recordings have been analysed and studies made of the thickness of fouling layers at major elevations of platforms. Early predictive analyses of marine growth have been found to be inaccurate. No distinctive differences in the fouling assemblages on steel platforms in the central and northern North Sea are apparent. Twenty-one sessile organisms are common and abundant on most platforms studied. In addition, five other species are abundant on a minority of platforms. The biology of these organisms is discussed and the present lack of information on their possible role in bringing about corrosion highlighted.

The Offshore Marine Studies Unit has been studying marine growth on oil and gas production platforms ranging in age from one year to eight years. These platforms are situated in twelve oilfields in the central and northern North Sea. The majority of them stand in water greater than 90 m deep and are more than 100 km from the Scottish coast; an important exception is Beatrice which stands in about 45 m of water and is situated about 20 km from the Scottish mainland.

Marine growth data are collected from scrape samples, colour photographs and video recordings. Scrape samples from each of the major elevations of a jacket are collected by divers whenever possible. Generally speaking, these elevations are situated at about −12 m, −30 m, −50 m, −70 m, −90 m and in deeper water at −120 m. At each elevation, two samples are taken from a leg, one from the outward facing surface and one from the inward facing surface, and two are taken from a position midway along a horizontal bracing, one from the upper surface and one from the lower surface. At each sample point a 30 cm × 30 cm quadrat, which is marked clearly in centimetres, is laid against the steel surface and a close-up colour photograph is taken of the marine growth *in situ*. A series of stand-off colour photographs are also taken at each sample point showing the fouling in the vicinity of the sample site. Marine growth is then scraped off from an area measuring 15 cm × 15 cm. Analysis of the samples and photographs is carried out in the laboratory to determine figures for percentage cover and average thickness of major fouling species. The data are supplemented by video recordings. Although we are able to identify the larger and more common organisms from black and white video recordings and estimate percentage cover, considerably more information can be gathered from colour video recordings.

The thickness of marine growth around a member is calculated from circumferential measurements made with a tape measure. We believe that this approach to estimating the increase in radius gives more meaningful results than spot measurements of individual organisms.

For example, many of the common seaweeds present on platforms are between 15 cm and 20 cm long but do not grow continuously all the way around members, being restricted to sunlit surfaces. Also these seaweeds, and many of the hydroids, die back in the winter. The tape measure compresses the soft growth giving a more representative reading of overall thickness. Likewise, circumferential measurements of hard growth are more meaningful; we have recorded considerable variability in thickness of hard growth using this simple method. For instance, mussel beds on horizontal bracings are never uniformly thick. Thickness measurements are important to both corrosion and strucural engineers.

Early appraisals of the patterns of marine growth on platforms in the central and northern North Sea predicted that jackets in the central North Sea would be more heavily and more rapidly fouled than those in the northern sector [1.2, 1]. This assumption was based on an evaluation of the current systems affecting the platforms. It was thought that coastal currents would be predominant around platforms in the central sector and oceanic currents would predominate in the northern sector. Coastal currents are known to transport large numbers of planktonic larvae and plant spores out into the North Sea from parent populations living along coasts.

Mussels were considered to be the most serious fouling organism likely to foul central North Sea platforms, whereas on the northern North Sea platforms kelps were expected to be of far greater importance [1.2, 1]. The mussels in the central sector of the North Sea were expected to show more rapid growth rates than those on platforms further north. The thickness of mussel beds in the depth range Mean Low Water (MLW) to −40 m on central North Sea platforms was anticipated to be at least 5 cm (2 in) within five years. In the northern sector, kelps were not expected to appear in significant numbers for several years but by the time the platforms were seven or eight years old these plants were thought likely to be a serious nuisance in the depth range MLW to −25 m. On the northern North Sea platforms, marine growth in general would be minimal for at least five years.

An analysis of the marine fouling data collected from platforms over the past three years shows that it is not possible to predict the fouling communities solely on the basis of the geographical location. Similarities exist on platforms in both sectors. For example, mussel populations on some platforms in the northern North Sea are as dense as those on platforms in the central North Sea. The growth of the mussels on platforms in both sectors is often very similar; we have recorded mussels 4 cm in length on northern North Sea platforms two years after the installation of the jacket. Indeed, on one northern North Sea platform mussels are already being cleaned off the horizontal bracings at the −40 ft elevation, only five years after installation. We also have evidence of slow growth rates of mussels in both sectors. Kelps are present on several platforms but we know of only two where these plants are sufficiently common to be a nuisance. Both have been installed for seven years, but are geographically widely separated: one is in the northern sector and one in the central sector.

It is apparent that coastal current systems are important and efficient conveyors of fouling organisms to platforms in both the northern and central sectors of the North Sea. Residual and wind blown currents may be of equal importance in this respect. Knight-Jones [1.2, 2] points out that larvae and spores transported in a current moving at only half a knot will be carried nearly 200 km in a week. Many larvae and spores are capable of delaying metamorphosis or germination for two or three months, increasing their chances of finding a suitable habitat for subsequent sessile life [1.2, 3], [1.2, 4], [1.2, 5]. Many larvae carried away from coasts by deep water currents are sufficiently strong swimmers to move vertically through the water column and to maintain station on coming into contact with a suitable substrate, ensuring settlement at the optimum depth range for the species [1.2, 6].

Residual and wind blown currents vary over a period of time. Such variations may play an important role in bringing about dissimilarities in fouling assemblages on platforms separated by relatively short distances. A recent publication by Forteath *et al.* [1.2, 7] highlights an example of dissimilar marine growth communities on neighbouring North Sea platforms. This paper compares the marine growth on the Montrose and Forties platforms. The Montrose platform is only 40 km to the south-east of the Forties field, stands in a similar depth of water as the Forties platforms and is of similar age. Major differences between the communities are set out below:

MLW to −30 m
MONTROSE. Seaweeds (upper stratum); arborescent bryozoans (lower stratum).
FORTIES. Mussels.

−30 m to −70 m
MONTROSE. Bryozoans. (a) Calcareous bryozoans dominate between −30 m and −50 m and (b) encrusting bryozoans dominate between −50 m and −70 m.
FORTIES. Anemones.

−70 m to the mud-line
MONTROSE. Aggregate tubeworms.
FORTIES. Aggregate tubeworms.

It is difficult to account for these differences without good scientific data on the biological, chemical and physical parameters of the sea in the proximity of each

Table 1.2, 1
Important fouling organisms on central and northern North Sea platforms

Seaweeds	Hydroids	Anthozoa	Bryozoa	Polychaetes	Bivalve molluscs	Crustacea
Enteromorpha intestinalis	*Obelia dichotoma*	*Metridium senile*	*Scruparia ambigua*	*Pomatoceros triqueter*	*Mytilus edulis*	†*Balanus hameri*
Ulva lactuca	*Laomedea flexuosa*	*Alcyonium digitatum*	*Electra pilosa*	*Hydroides norvegica*	*Pododesmus patelliformis*	**B. balanus*
Polysiphonia urceolata	*Tubularia larynx*		**Alcyonidium hirsutum*	†*Filograna implexa*		**B. crenatus*
P. brodiaei	†*T. coronata*		*Omalosecosa ramulosa*			
**Laminaria digitata*	†*Bougainvillia muscoides*					
**Alaria esculenta*	*B. ramosa*					

*Important on a minority of platforms.
†Most abundant in the deep-water regions of platforms.

jacket at the time of installation, and in subsequent years. The history of a platform clearly is relevant to the present state of fouling on it and the present state of fouling cannot be adequately explained without specific knowledge of historical events.

About 200 species of seaweeds and invertebrates have been recorded on North Sea platforms. Table 1.2, 1 lists 26 species which are either present on all platforms studied or dominate a certain depth range on one or two platforms. It is interesting that only four species are deep-water organisms and that the remainder are either intertidal or sublittoral coastal species.

Pipe [1.2, 8] has outlined the manner in which fouling organisms may bring about corrosion. Subsequent papers in this chapter will be concerned with the biological mechanisms leading to potential corrosion problems. Below are outlined relevant aspects of the biology, distribution and abundance of the more common fouling organisms.

SEAWEEDS

The four most common seaweeds on jackets in the central and northern North Sea are the green seaweeds *Enteromorpha intestinalis*, and sea lettuce *Ulva lactuca*, and the red seaweeds *Polysiphonia urceolata*, and *P. brodiaei*. *E. intestinalis* is common in the splash zone on nearly all North Sea production platforms. The rhizoids of this species are known to penetrate surface levels of paint and cause extensive blistering and cracking [1.2, 9]. Water also is retained beneath the fronds during periods of low water, creating extremely corrosive conditions. Several operators have expressed their concern at the problems of protecting the steel surfaces in the splash zone. *U. lactuca* is common in the depth range MLW to about −6 m. Dense carpets of this seaweed are present on many platforms, particularly on legs and diagonal bracings. Below −6 m to about −12 m, the red seaweeds dominate. In the absence of mussels, these four species of seaweed will more or less completely cover the underlying steel on sunlit surfaces of a platform. On Montrose, *U. lactuca* and *P. urceolata* completely cover large areas of the upper surface of horizontal bracings at the −10 m elevation [1.2,7]. A dense growth of *P. urceolata* and *P. brodiaei* is present also on horizontal members at −12 m on a platform to the North of Shetland. All of these species are able to reproduce vegetatively; prostrate axes attach at intervals to the substrate and new plants grow up at these attachment points [1.2,10]. This mode of reproduction provides a means for considerable penetration and undoubtedly has contributed to the success of these species on North Sea installations. It also means that rhizoid penetration of protective coatings will take place over a considerable area of the steel surface. During the summer, silt and debris will become trapped in the filamentatous fronds which may favour the development of bacterial populations, resulting in severe oxygen depletion and possibly anaerobic corrosion processes. However, these seaweeds die back to small root-like processes during winter months and much of the silt will be washed away.

Kelps and the kelp-like *Alaria* are not widespread on platforms in the North Sea but where they do occur, numbers are increasing annually. Examination of holdfasts pulled away with the plant show small pieces of metal trapped amongst them and in some instances, quite large pieces of protective coating. Kelps are quite simple to remove. An extensive kelp forest has recently been destroyed by severing each plant with a knife, the divers completing the operation in about twelve hours. We hope to examine colour videos of the cleaned area and observe the extent of break-down of stipes and holdfasts. Decomposition of this organic material may well bring about corrosion at the steel/holdfast interface.

HYDROIDS AND BRYOZOANS

Hydroids and bryozoans are very common on North Sea platforms. The smaller species of hydroids, especially *Obelia dichotoma, Laomedea flexuosa* and *Bougainvillia ramosa* do not in themselves present a nuisance since they are usually less than 5 cm in length and seldom sufficiently dense to trap silt and debris. In the winter these species of hydroid die or enter a resting phase. However, the hydrocauli of these species provide mechanical support for colonies of the bryozoans *Scruparia ambigua, Electra pilosa* and *Alcyonidium hirsutum.* A hydroid/bryozoan 'turf' can become established rapidly and cover 100 per cent of the underlying steel over extensive areas of jacket. This turf ranges in thickness from a few millimetres to as much as 5 cm. The turf is more or less permanent since the bryozoans continue to grow throughout the winter and may live for several years. Between MLW and −12 m, the hydroid/bryozoan turf is usually confined to the shaded surfaces of legs, diagonal and horizontal members: sunlit surfaces are almost invariably covered by seaweeds. Below −12 m the hydroid/bryozoan turf dominates all surfaces. In the depth range MLW to about −30 m, *E. pilosa* is extremely common on Montrose and several other platforms, forming extensive carpets 4-5 cm thick. In the mid-water depth range (about −50 m to −70 m) *A. hirsutum* is often common, encrusting in particular the hydrocauli ('stem') of *O. dichotoma* and *B. ramosa*. However, to date *A. hirsutum* has been found to dominate the marine growth assemblage in this depth range only on Montrose. The calcareous bryozoan *Omalosecosa ramulosa* is also common in the mid-water depth range of the majority of jackets we have studied but is confined largely to vertical surfaces, particularly legs and risers.

Tubularia larynx and *T. coronata* are the largest species of hydroid colonising central and northern North Sea oil platforms. (We have not recorded *T. indivisa* although it is very common on platforms in the southern North Sea.) Colonies of *T. larynx* are between 6 cm and 10 cm high and are most common in the depth range MLW to about −50 m. *T. coronata* colonies are between 10 cm and 15 cm high and are most common in the deep-water regions of platforms. These two species often dominate the fouling community for the first two or three years of a jacket's life, and thereafter numbers of colonies rapidly decline. We have found large accumulations of silt and debris trapped in the root-like attachments of both species. Samples scraped off the steel surface often have corrosion products and protective coating adhering to them. Both species have been found to colonise cathodic protection anodes and in several instances the anode has been completely obscured by a mat of *Tubularia*.

The deep-water hydroid *Bougainvillia muscoides* is very common on most platforms that we have studied, particularly below −60/70 m. Four or five years after the installation of a jacket this species may cover more or less 100 per cent of the available horizontal surfaces in the deep-water depth range. Although only 5 or 6 cm high, the hydroid grows in an amazingly profuse manner, rapidly covering large areas of the substrate in a tangled matrix of fused and frequently branching stolons. Large accumulations of silt become trapped in this matrix. The species is present all year round and shows no sign of die-back during winter months. We have found that this hydroid is able to tolerate considerable amounts of silt and is often the only organism living on the upper surfaces of horizontal members which are badly affected by drill cuttings. This hydroid is also very abundant on risers in the deep-water depth range of jackets. Ramussen [1.2, 11] recorded dense colonies fouling the intake of a power station in the Isefjord but little is known about the biology of this potentially important fouling organism. The medusa stage is very inactive and the breeding season appears to be more or less confined to the autumn; at this time of the year deep water reaches its annual maximum temperature and possibly stimulates reproduction [1.2, 12].

ANEMONES AND SOFT CORALS

The most common anemone on North Sea platforms is the plumose anemone *Metridium senile*, which is present on platforms in the southern, central and northern North Sea. On central and northern North Sea platforms it has been recorded at all depths but is usually most abundant in the mid-water depth zone (about −30 m to about −70 m) where it shows preference for horizontal and diagonal surfaces. On the much smaller southern North Sea platforms *M. senile* is abundant from the surface to the mud-line and covers vertical, diagonal and horizontal bracings.

Asexual reproduction is a special feature of the life cycle of *Metridium senile* [1.2, 13]. A single sexually produced planula larva is capable of eventually giving rise to many hundreds of asexually produced offspring by a process of pedal laceration, [1.2, 14], [1.2, 15]. The clonemates from an asexually reproducing individual are remarkably alike and can be distinguished easily from members of other clones by body colour [1.2, 16]. On some southern North Sea platforms, *Metridium* has covered 60 per cent of the jacket within four or five years. Recolonisation after cleaning is more rapid. Many individuals have been seen to climb back up the jacket by divers soon after removal. On platforms in the central and northern sectors, colonisation has not been as fast. However, on Forties large numbers of clones have now become established [1.2, 17]. On the four platforms in this oilfield, it has taken *Metridium* six years to dominate the fouling community of the mid-water region.

Metridium senile exhibits inter-clonal aggression [1.2, 16]. The peripheral members of a clone develop specialised stinging cells which may prevent integration of clones [1.2,18]. On platforms where large numbers of clones have become established or clones are aggressive towards one another the individual anemones become sessile. Sessile individuals will obscure the underlying steel preventing routine visual inspection for pitting or other signs of corrosion. A study of the steel/anemone interface would be interesting; silt and faeces presumably will collect beneath and between individuals and may lead to corrosion of the underlying steel.

The soft coral, *Alcyonium digitatum*, is also widespread and common on platforms in the southern, central and northern North Sea. On platforms in the central and northern sectors, colonies are present at all elevations. Colonies appear to be most numerous on vertical bracings, pile guides and gusset plates. On some platforms in the southern gasfields, *A. digitatum* is very abundant and large areas of legs and other vertical surfaces are completely covered by numerous colonies at least 20 cm high. On the majority of platforms in the central and northern North Sea, colonies are more widely distributed, the more dense patches seldom covering more than 30 cm^2 or 40 cm^2. However, on a few jackets, colonies are extensive and cover large areas. The basal diameter of some of these colonies measures 8 cm. Colonies adhere strongly to the steel and basal portions are not easily removed. It is not known whether pitting takes place beneath the colonies but corrosion products have been found attached to the basal areas of colonies. Growing colonies will trap sediments and debris creating pockets of anaerobic material. Colonies may live for several years and in adverse conditions they may enter into a resting stage [1.2, 19] resulting in areas of corrosion being hidden from view for a long period of time.

POLYCHAETES

The serpulids *Pomatoceros triqueter, Hydroides norvegica* and *Filograna implexa* are present on all the platforms we have studied in the central and northern North Sea. The solitary tubeworms *P. triqueter* and *H. norvegica* are 'opportunist' species and are usually amongst the first fouling organisms to be recorded on platforms after installation and cleaning. Both species breed in the summer and therefore larvae are present in the sea at the time most jackets are installed or are undergoing cleaning. The aggregate tubeworm *F. implexa* probably has the widest distribution of all the serpulid worms [1.2, 20].

The solitary tubes of *Pomatoceros* and *Hydroides* are usually scattered over all surfaces from MLW to the mud-line. After three or four years most of these animals become smothered by colonial organisms and seaweeds. Mono-specific dominance by this species is very uncommon [1.2, 21]. However, *P. triqueter* forms dense aggregations on one platform close to the Scottish mainland and dominates the fouling community over extensive areas. On this one platform, the feeding apertures of *P. triqueter* are atypically directed upwards from the steel substrate and the calcareous tubes form a continuous fouling layer about 2 cm thick. Dense aggregations of tube-dwelling polychaetes are less susceptible to overgrowth than isolated individuals of the same species [1.2, 22]. The tightly packed tubes of *P. triqueter* on this platform would present an outer 'colony' surface consisting almost entirely of feeding apertures which would reduce the probability of successful settlement by larvae of other species [1.2, 23]. However, individuals only live for between 1 and 1½ years [1.2, 28] and the present beds may become colonised by other organisms over the next few years: the platform is only two years old. On some of the southern gas platforms even the most extensive aggregations of solitary tubeworms have been overgrown by hydroids, anemones and soft corals within two or three years. A dense fouling layer is likely to become estabilshed on platforms where aggregations of solitary tubeworms occur. This layer may reduce corrosion of the underlying surface as a result of the complete or partial exclusion of oxygen [1.2, 24].

The aggregate tubeworm *Filograna implexa* is most common on surfaces in the deep-water regions of platforms (about −60 m to the mud-line). Aggregations of this worm are present on vertical surfaces and surfaces more or less free of silt; very few aggregations have been recorded on the upper surfaces of horizontal and diagonal members, which are often affected by cuttings. Initially, the worms secrete very small tubes which encrust the substrate. One suspects that relatively small amounts of drilling mud and silt smother them and this would appear to be substantiated by their distribution. Competition for feeding space results in the formation of dome-shaped pseudocolonies [1.2, 25]. Pseudocolonies on North Sea platforms may have a diameter of 25–35 cm and height of 25-30 cm. These structures often coalesce forming a reef over extensive areas but, unlike coral, the tubes are fragile and easily broken. Divers have little problem removing *Filograna* from steel surfaces.

BIVALVE MOLLUSCS

Common mussels, *Mytilus edulis*, and saddle oysters, *Pododesmus patelliformis*, are present on every platform we have studied. *M. edulis* settles in the depth range MLW to about −25 m but *P. patelliformis* settles at all elevations of the jacket. Both species attach themselves to the substrate by means of byssus threads. Mussels secrete large numbers of byssus threads and where beds of these animals occur, a tangled matrix of threads is produced. Saddle oysters disperse more widely over the substrate and seldom produce more than one or two byssus threads, which hold one valve of the animals's shell flush to the steel surface.

Nearly all the steel structures we have examined in the central and northern North Sea have a major elevation at about −12 m. The horizontal bracings at this elevation are almost invariably colonised by mussels. On some platforms in both the central and northern North Sea, mussels form dense and extensive beds over the steel surface of these bracings. On other platforms, surfaces are covered by a complex fouling assemblage and mussels form relatively small patches. Mono-specific beds of mussels on North Sea platforms are made up of fast-growing individuals, the majority of which reach a length of 4 cm during their second summer and about 8 cm in their eighth summer. Mussels present in a mixed fouling assemblage grow at a much slower rate: many mussels will be less than 2.5 cm long at the age of three and less than 4.5 cm long at the age of five. Mussels in a mixed fouling assemblage on a platform have to compete for food with several other filter feeding organisms such as hydroids and also may be handicapped during feeding by the whiplash effect of seaweeds, which are waved back and forth continually by waves and currents.

Mussel beds on platforms accumulate a considerable amount of organic debris which becomes trapped in the tangled mass of byssus threads. The depth and extent of these accumulations will vary considerably. Mussel beds are most dense around nodes and the support members of anodes and become progressively less dense towards the centre of bracings. At locations where mussels are most dense, anaerobic conditions probably prevail at the steel/mussel bed interface but towards the centre of bracings some oxygen almost certainly will be present. We have no knowledge of any studies on the corrosion effects of mussel beds on steel structures in the North Sea but such a study would be interesting in view of the increasing number of platforms becoming fouled by this organism.

On some platforms, mussels completely cover some anodes, particularly those attached to horizontals at the −12 m elevation. The extent to which this inhibits the efficient functioning of these anodes is not known but wastage probably is minimal because the mussels remain attached to the surface of these anodes and often measure several centimetres in length.

Light cleaning of mussels is relatively simple using water jetting. However, light cleaning invariably leaves large numbers of byssus threads attached to the steel which may become encrusted by other fouling organisms, especially *Electra pilosa*, or continue to collect silt. The effective removal of byssus threads requires wire brushing or grit blasting and on structures which have been painted with a protective coating this may cause considerable damage.

Saddle oysters are opportunists, and together with solitary tubeworms, are amongst the first organisms to settle on newly installed or cleaned structures. Specimens on North Sea platforms are smaller than those living inshore; the largest specimens we have recorded are about 1 cm in diameter. However, these animals are overgrown and smothered by colonial species within two or three years. Dead shells will remain attached to the steel for several years but unlike the baseplates of barnacles are easily removed.

CRUSTACEA

Barnacles found on central and northern North Sea platforms are sublittoral species. In the shallow regions, *Balanus crenatus* and *B. balanus* are the two species most consistently recorded. *B. balanus* is present also in mid-water regions on some platforms. Below a depth of about −60 m, the deep-water barnacle *Balanus hameri* has been recorded on nearly every platform studied which suggests that this species is much more common than previously believed.

Balanus crenatus is abundant on only one platform studied. This structure lies close to the Scottish mainland and has been installed for three years. Large numbers of *B. crenatus* cyprids must have settled on this structure during the first spring after installation resulting in a mono-specific fouling layer over an extensive area. The barnacles are so densely packed along the members that they are enormously elongated. The majority of individuals are 2 cm high. [1.2, 26] and Knight-Jones and Moyse [1.2, 25] describe similar growth patterns following mass settlement of *B. crenatus* cyprids on test panels and rocky shores.

Belanus crenatus is a short-lived species and by the autumn of 1981 the majority of the individuals on the platform we studied were dead. The beds were already becoming overgrown by sea-squirts, soft corals and hydroids. It is probable that the beds will break up in the next year or two. Barnes and Powell [1.2, 26] found crowded populations of *B. crenatus* easy to detach since the calcareous parts were extremely fragile and attachment area feeble.

The deep-water *Balanus hameri* possesses a calcareous baseplate and divers have great difficulty removing them. Baseplates remain attached to the steel even after rigorous cleaning. Severe localised pitting of steel is known to take place under barnacles [1.2, 8] and there are reports of barnacles gouging their way through 800 μm of coal tar coating per year [1.2, 27]. Neither species is particularly abundant on the platforms we have studied, and many become overgrown and smothered by hydroids and encrusting bryozoans two or three years after installation.

CONCLUSION

It is not possible to divide the marine growth on platforms in the central and northern North Sea into two distinctive community types. It has been shown that even neighbouring platforms may have significantly different fouling communities. This paper has identified those organisms which are widely distributed on platforms in both sectors of the North Sea and also organisms which are sufficiently common on a minority of platforms to warrant special mention. On platforms in both sectors, growth rates, abundance and distribution of many of these organisms appear to be similar. It is becoming possible now to predict with increasing accuracy the type of fouling community developing on each of the platforms studied once the initial macrofouling community on a given platform has been identified.

Corrosion products have been found attached to a variety of marine organisms collected from platforms. However, it is not known if the corrosion is a direct result of conditions prevailing beneath the fouling layer. Corrosion may have taken place soon after the structure was installed but remained undetected because of the speed at which fouling obscures the steel surface on platforms in the central and northern North Sea. Corrosion beneath soft growth may remain undetected for a longer period of time than that beneath hard growth. The cleaning of mussel beds appears to be necessary after about 6 or 8 years but the cleaning of areas covered by soft growth may not take place for at least 10 or 12 years. Annual marine growth inspections may save the operator considerable costs by revealing areas of corrosion which otherwise might go undetected for many years.

REFERENCES

[1.2,1] Oldfield, D. G. Appraisal of Marine Fouling on Offshore Structures, *CIRIA*, London, 1978.

[1.2,2] Knight-Jones, E.W. *J. Mar. Biol. Ass. U.K.*, 1951, **30**, 201-207.

[1.2,3] Thorson, G. *Biological Reviews*, 1950, **25**, 1-45.

[1.2,4] Wilson, D. P. *Ann. Inst. Oceangr.* Monaco, 1952, **27**, 49-156.

[1.2,5] Bayne, B. L. *Ophelia*, 1965, **2**, 1-47.

[1.2,6] Mileikovsky, S. A. *Mar. Biol.*, 1973, **23**, 11-17.

[1.2,7] Forteath, G. N. R., Picken, G. B., Ralph, R. and Williams, J. *Mar. Ecol. Pro. Ser.*, 1982, **8**, 61-68.

[1.2,8] Pipe, A. North Sea fouling organisms and their potential effects on the corrosion of North Sea structures. In *Symposium on Marine Corrosion on Offshore Structures*. Society of Chemical Industry's Aberdeen Section and Materials Preservation Group, University of Aberdeen, 1981.

[1.2,9] Terry, L. A. and Endyvean, R. G. J. *Botanica Marina*, 1981, **24**, 177-183.

[1.2,10] Fritsch, F. E. *The Structure and Reproduction of the Algae*. Vol. 2. Cambridge University Press, London, 1945.

[1.2,11] Ramussen, E. *Ophelia*, 1973, **11**, 1-507.

[1.2,12] Edwards, C. *J. Mar. Biol. Ass. U.K.*, 1964, **44**, 725-752.

[1.2,13] Hoffman, R. J. *Biol. Bull.*, 1976, **151**, 478-488.

[1.2, 14] Stephenson, T. A. *The British Sea Anemones,* Vol. II, The Royal Society, London, 1935.

[1.2, 15] Gemmill, J. F. *Phil. Trans. Roy. Soc., London Ser. B.*, 1970, **209**, 351-375.

[1.2, 16] Purcell, J. E. *Biol. Bull.*, 1977, **153**, 355-368.

[1.2, 17] Goodman, K. S, and Ralph, R. Animal fouling on the Forties platform. In *Marine Fouling of Offshore Structures*. Society for Underwater Technology, London, 19-20 May 1981.

[1.2, 18] Purcell, J. E. and Kitting, C. L. *Biol. Bull.*, 1982, **162** (3), 345-358.

[1.2, 19] Hartnell, R. G. Reproduction strategy in two British species of Alcyonium. In *Biology of Benthic Organisms,* edited by Keagen, B. F. *et al.* Pergamon Press, Oxford, 1976.

[1.2, 20] Hove, H. A. Different causes of mass occurrence in serpulids. In *Biology and Systematics of Colonial Organisms,* edited by Larwood, G. and Rosen, B. R. Academic Press, London, 1979.

[1.2, 21] Lewis, J. R. *The Ecology of Rocky Shores*. Hodder and Stoughton, London, 1964.

[1.2, 22] Jackson, J. B. C. *Amer. Nat.*, 1977, **111**, 743-767.

[1.2, 23] Wilson, D. P. *J. Mar. Biol. Ass. U.K.*, 1968, **48**, 387-435.

[1.2, 24] Southwell, G. R., Bultman, J. D. and Humer, C. W. Influence of marine organisms on the life of structural steels in sea water. Naval Research Laboratory, Washington, DC, Report 7672, 1974.

[1.2, 25] Knight-Jones, E. W. and Moyse, J. Intraspecific competition in sedentary and marine animals. In *Symposium of the Society for Experimental Biology,* 1961, **15**, 72-95.

[1.2, 26] Barnes, H. and Powell, H. T. *J. Anim. Ecol.*, 1950, **19**, 175-179.

[1.2, 27] Kirkwood, D. Marine fouling and corrosion. In *Marine Fouling of Offshore Structures*. Society for Underwater Technology, London, 1981.

[1.2, 28] Nair, N. B. Ecology of marine fouling and wood-boring organisms of Western Norway, *Sarsia*, 1962, **8**, 1-88.

1.3 SULPHATE-REDUCING BACTERIA AND ANAEROBIC CORROSION

W. A. HAMILTON and P. F. SANDERS, Department of Microbiology, University of Aberdeen, Aberdeen, Scotland

ABSTRACT

While there is controversy over the detailed mechanism of anaerobic corrosion, what is not in doubt is the direct involvement of the sulphate-reducing bacteria. The role of sulphate-reducing bacteria is ultimately dependent upon three features of their physiology; they are anaerobes, they have a restricted nutritional spectrum, they produce sulphide. The first two characteristics determine the environments in which they are active, while the third is both the root cause of their problem potential and the means whereby we can gain a measure of that potential. Marine sediments and the macrofouling associated with offshore structures provide conditions in which sulphate-reducing bacteria could be active, and where that activity could cause corrosion problems.

The question is, how can we assess the practical importance of these effects?

INTRODUCTION

Sulphate-reducing bacteria are implicated in a wide range of problems experienced by the oil and gas industries. These may be grouped under two headings, only the second of which is peculiar to the offshore environment.

Problems in which sulphate-reducing bacteria are implicated

Production Systems

Reservoir	: souring, plugging
Oil-carrying lines (risers, pipelines)	: internal corrosion
Water-carrying lines (injection systems)	: internal corrosion
Storage	: H_2S gas, corrosion, souring

Offshore Structures

Marine macrofouling Natural sediments Deposited materials	} external corrosion of underlying structure

FUNDAMENTAL QUESTIONS

The one feature common to all these phenomena is the production of sulphide, and this immediately raises two fundamental questions.

(a) What is the biological nature of this sulphide production?
(b) Why does it arise in these particular environments?

The sulphate-reducing bacteria are strict anaerobes. That is, they can only grow and demonstrate biological activity in the complete absence of oxygen. (They can, however, be cultured from aerobic environments such as sea-water, where they will normally exist in a dormant biologically inactive form.) In a manner directly analogous to the reduction of oxygen by normal respiratory organisms, sulphate-reducing bacteria require sulphate as a terminal electron acceptor and their biological activity is therefore inevitably associated with the production of sulphide.

Environments in which active sulphate-reducing bacteria are to be found are therefore depleted of oxygen and rich in sulphate. The nature of the oxygen depletion is particularly interesting. Most often it depends upon the net effect of a restricted access to oxygen, and the activity of aerobic respiratory organisms which utilise the available oxygen, while at the same time chemically modifying the environment by their growth on the primary substrate(s). This latter property is also highly significant for the sulphate-reducing bacteria which are relatively restricted with regard to their own nutritional spectrum; they cannot, for example, grow on hydrocarbon or carbohydrate materials. What they can do, and do very well, is grow on compounds such as alcohols, fatty acids and hydrogen which are formed as partial breakdown products of the metabolic activities of the oxygen-utilising species. Sulphate-reducing bacteria are therefore always found in close association with other biological species which play key roles in the creation of the necessary physical and chemical conditions for their growth and attendant sulphide production.

PRACTICAL QUESTIONS

On offshore production structures it is likely that such environmental conditions might obtain under marine macrofouling, natural marine sediments or deposited materials such as discarded cuttings. These particular environments are therefore likely to harbour active populations of sulphate-reducing bacteria with the consequent potential for corrosion of the underlying steel structure. Although uncertainty remains over the significance of every component reaction in the complex mechanism of anaerobic corrosion, the capabilities of sulphate-reducing bacteria (a) to cause cathodic depolarisation by utilisation of the electrochemically produced hydrogen, and (b) to be instrumental in the continuous production of sulphide and the deposition of sulphide films, clearly identify these organisms as playing a critical role in the overall process.

These considerations thus lead to three important practical questions. What is the nature of any anaerobic corrosion associated with, for example, marine macrofouling; in respect of frequency of occurrence, extent, location, nature of substrate, etc.? Is there a direct relationship between biological activity (as discussed above) and corrosion activity (as measured by weight loss, pitting, identification of corrosion products)? What experimental methods and protocols must be developed in order to obtain answers to these first two questions? This last point is central to any analyses of corrosion processes with regard to quantification, prediction and assessment of the efficacy of any preventative measures.

METHODOLOGY

Methodology is best considered under two separate subheadings; sample collection and handling, laboratory assay.

Sample collection and handling

The microscopic size of the sulphate-reducing and other bacteria, and often their very close physical association with the steel surface, raise particular problems with regard to observation and sampling. Television pictures from RCVs or divers' visual and photographic reports can give no relevant microbiological information, beyond any evidence of pitting corrosion *per se* or the appearance of black sulphide films. While scrape samples from the steel surface or grab samples from overlying sedimented material are of some limited value, critical analyses are dependent on the recovery of undisturbed metal coupons or sediment core samples. Such material must then be transferred to a laboratory before the signal can be amplified and the relevant microbiological data obtained. Ideally this laboratory stage of the analyses should be at least initiated offshore, and in any event sample material must be rigorously protected against all forms of stress; for example, temperature fluctuation, exposure to air, drying out, mechanical shock, mixing of sediment material.

Laboratory assay

The microbiological parameters of any given environment are classically determined by culturing the various organisms present, using a range of selective media to allow the growth and separate identification of the individual species comprising the resident populations. These techniques are routinely used within the oil industry to gain information on the presence and numbers of sulphate-reducing and other bacteria associated with a specific problem, or potential problem. For a number of technical reasons, however, the information obtained is only, at best, an indirect measure of the microbiological activity of the system under study. Most fundamentally, it tells the experimenter what are certain of the capabilities of the organisms from the environment, as expressed and measured under one particular set of tightly defined laboratory conditions; this does not necessarily relate either qualitatively or quantitatively to the activity of the microbial population as a whole in the environment in question, and under the physical and chemical conditions existing in that environment.

The most straightforward way of obtaining such information is by direct analysis of the turnover of a key substrate under conditions as closely as possible reproducing those of the natural environment. In the case of the sulphate-reducing bacteria, the substrate and method of choice is the turnover of ^{35}S-radioactively labelled sulphate. Not only does it give a direct measure of the *in situ* biological activity of the organisms, but the assay is both sensitive and reproducible, and can give quantitative data within 36 h.

Sulphate reduction assay

The sample, which may be a corrosion coupon (3 × 1.5 cm), scrape sample, or core subsample, is placed in a test-tube (Fig. 1.3,1). The activity of the complete biological community associated with the sample is thus measured. Filtered sea-water containing a small amount of ^{35}S-labelled sodium sulphate is added, the tube flushed with oxygen-free nitrogen and sealed with a neoprene rubber stopper from which is suspended the cup and folded filter paper of the sulphide-trapping mechanism. Incubation is for 24 h at the *in situ* temperature. 1 ml of 1 M zinc acetate is then injected through the stopper on to the filter paper, and the reaction stopped by addition to the coupon/sea water of 1 ml 6 N HCl + 37% (w/v) formalin, 1:1. Acid-labile sulphides are released as H_2S and during an equilibration period of 12 h are trapped on the zinc acetate-loaded filter paper.

All the above manipulations should be carried out offshore, and immediately following retrieval of the sample. Any period of delay can seriously affect the accuracy of the assay.

Onshore, the filter paper wicks are transferred to a liquid scintillation radioactive counter and the level of ^{35}S-labelled sulphide determined. The rate of sulphate reduction is then readily calculated from the equation:

$$\text{Rate of } SO_4^{=} \text{ reduction (n moles/sample/day)} = \frac{\text{DPM}\,^{35}S^{2-} \text{ on wick}}{\text{DPM}\,^{35}SO_4^{2-} \text{ added}} \times 1.06 \times \frac{\text{amount } SO_4^{2-} \text{ added (n moles)}}{\text{time of incubation (days)}}$$

where 1.06 is the factor to allow for the differential partition of ^{34}S and ^{35}S due to microbial activity. (DPM = distintegrations per minute.)

This method has a number of marked advantages. It is relatively easy to perform, either as a single assay or in a programme of multiple samples; it is sensitive, reproducible and can give an answer within 2 days. Much the most important point, however, is that it gives a direct measure of the most significant activity of the biological community as a whole.

The central features of our work and of this presentation, are the assessment of the applicability of this activity assay to offshore environments, and the establishment of the nature of the relationship between biological and corrosive activities.

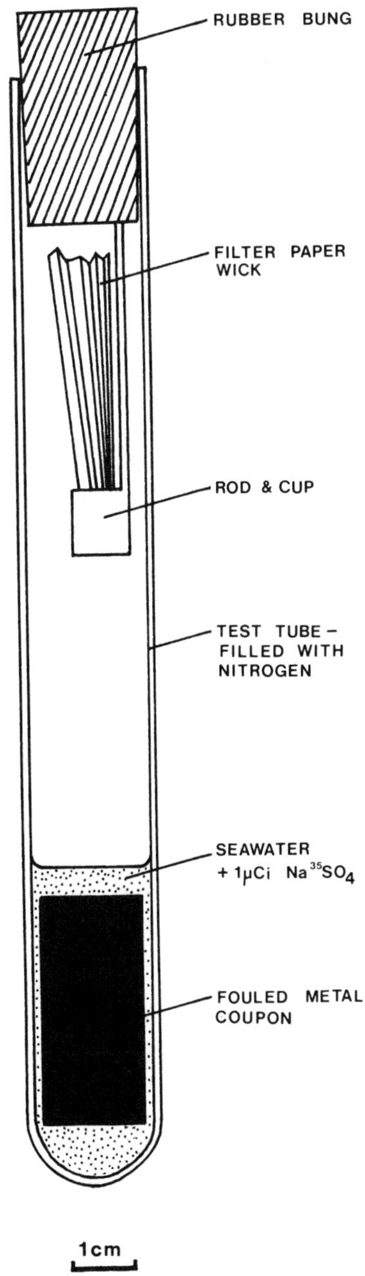

Fig. 1.3, 1 – Sulphate reduction assay.

EXPERIMENTAL FINDINGS

Test systems

Two onshore test systems have been used in the initial appraisal of the methodology in a field situation; the River Don estuary and Stonehaven harbour.

The estuary of the River Don carries a cellulose-based pollution load and its sediments are rich in sulphate-reducing bacteria. Corrosion coupons were sited in the water column and samples withdrawn for examination after 4, 11, 16 and 22 weeks during the spring and early summer. During this period some degree of soft algal fouling developed. Coupons of five different compositions were employed; 43 A, 50 D and Corten (0.49% Cu) steels, and the high-Cu alloys, Cu-Ni (90/10) and Monel 400. Analyses involved: removal of organisms by sonication, culturing on media selective for aerobic, anaerobic and sulphate-reducing species, and direct microscopic total count; assay of sulphate reduction activity; determination of corrosion by weight loss.

Due to removal of Cu by sonication and carry-over into growth media, no organisms could be cultured from Cu–Ni; it must be stressed that this negative result is an artefact resulting from the experimental techniques employed. The other data indicated the rapid establishment within 4 to 11 weeks of bacterial populations which remained relatively constant thereafter; total counts were 10^8-10^9 organisms/coupon, with sulphate reducing bacteria of the order of 10^2 organisms/coupon.

Only after 22 weeks was there the development of significant sulphate reduction activity. This activity was zero on Cu–Ni and Monel, fairly high on 43 A and Corten, and noticeably higher on 50 D. These data therefore differed qualitatively from the information on bacterial numbers, but bore a striking similarity to the findings with respect to weight loss.

The three steels showed significant corrosion rates from the earliest sampling time (4 weeks), and this countinued to the end of the period, with evidence latterly for a slight decrease in the rate of weight loss. Although analysis of corrosion products was not carried out, it is likely that this represents normal aerobic corrosion or rusting. With 50 D, however, there was a fourfold increase in the rate of weight loss at 22 weeks which corresponded exactly with the establishment of the high rate of sulphate reduction.

Stonehaven harbour, by comparison, represents a 'clean' site in terms of the lack of a major source of sulphate-reducing bacteria. Exposure was for a period of 18 months after which coupons had 80 to 90 per cent cover of soft fouling. Bacterial numbers were similar to those found on coupons from the Don estuary, but both the sulphate reduction rate and the weight loss were considerably lower.

These studies therefore established that the methods we were employing were capable of giving sensible field data, and furthermore they suggested that there might be an important correlation between biological suphate reduction activity and the corrosion weight loss of test coupons.

Offshore structures

With the cooperation of several operating companies, a number of sets of corrosion coupons have been sited on and close to various production structures in the cental and northern North Sea. The placement of such coupons, and their controlled retrieval over the next two or three years, will be essential to the establishment of a proper data base concerning the occurrence and nature of anaerobic bacterial corrosion problems affecting the offshore oil industry.

To date, preliminary findings only are available from coupons sited in the Fulmar and Brent fields. The data are given in Tables 1.3,1 and 1.3,2 and are reproduced here with the permission of Shell UK Production and Development Ltd. Coupons in Fulmar were suspended at two depths from the moored tanker. Exposure was for a 4-month period (Feb-May) during which no significant macrofouling developed. At Brent the coupons were also some distance from the production platform being suspended from a bouy to a single depth. Exposure was again over a 4-month period, but in May to August, and a macrofouling community of hydroids developed. In both cases the programme was restricted to Cu-Ni and 50 D, the latter with and without cathodic protection (CP) from the attachment of zinc sacrificial anodes.

Remarkably similar findings were obtained from both sites, and these confirmed and extended the suggestive data coming from the onshore test systems. No bacteria could be cultured from the Cu-Ni coupons, and the

Table 1.3,1

Metal coupons exposed for 113 days: Fulmar field

	Fouling: none visible	−5 m level	−20 m level
$SO_4^{=}$ reduction (pmole coupon^{-1} day^{-1})	Cu/Ni	385	683
	50 D	16 100	5250
	50 D + CP	Below detection	Below detection
Bacterial counts (No./coupon)	Cu/Ni	No growth	No growth
	50 D aerobic	2×10^5	Not determined
	50 D SRB	5×10^2	Not determined
	50 D + CP aerobic	1×10^7	1×10^6
	50 D + CP SRB	3×10^2	4×10^2
Corrosion rates (mdd)	Cu/Ni	7.7	–
	50 D	502.2	–
	50 D + CP	15.7	–
	Sulphate concentration: 2.2 g l^{-1}		

Table 1.3,2

Metal coupons exposed for 111 days: Brent field

Fouling: 100 per cent *Obelia dichotoma* on 50 D steel			
Sulphate turnover (pmole/coupon/day):			
	Cu/Ni	3 610	
	50 D	13 120	
	50 D + CP	230	
Bacterial numbers (No./coupon):			
	50 D	Aerobes	4×10^6
		SRB	2×10^2
	50 D + CP	Aerobes	2×10^6
		SRB	9×10^1
Corrosion rates (mdd; μm/year):			
	Cu/Ni	15.5; 67.4	
	50 D	35.2; 165	
	50 D + CP	nil	

numbers from 50 D were not affected by either depth or CP. The pattern for both sulphate reduction and corrosion weight loss differed markedly from this however; low on Cu-Ni, higher on 50 D, drastically reduced by CP.

CONCLUSION

While it must be stressed that these are indeed only preliminary findings, they do hold considerable promise that the sulphate reduction activity assay will prove to be a rapid, accurate and direct measure of that critical biological activity which is, in turn the direct cause of the anaerobic corrosion of ferrous metals. We plan to confirm and extend this proposal, and to establish the technique as a quantitative method of predicting the course of any likely corrosion problem, and of assessing the effectiveness of preventative measures employed.

ACKNOWLEDGEMENTS

We should like to express our indebtedness to our colleagues Drs H. R. Rosser and J. A. Hardy whose earlier work established the foundations on which this report is based. The work of our laboratory is supported by the Marine Technology Directorate of the Science and Engineering Research Council, and by finance and collaborative projects with BP Petroleum Development (UK) Ltd, and Shell UK Production and Development Ltd.

1.4 BIOLOGICAL MECHANISMS LEADING TO POTENTIAL CORROSION PROBLEMS

T. G. WILKINSON, Shell UK Exploration and Production Ltd, Aberdeen, Scotland

ABSTRACT

Biological/environmental problems occurring from offshore oil operations are often associated with damage to the environment. However, the biological reactions can also have severe effects on offshore operations.

Experience offshore has shown that significant concentrations of hydrogen sulphide are generated by sulphate-producing bacteria (SRB) in stagnant water in offshore installations.

The formation of stagnant conditions results from the growth of aerobic bacteria in large volumes of water in, for example, either the legs or storage cells of offshore structures. Once the oxygen is consumed, conditions are ideal for the growth of SRB which can use the sulphate present in sea-water for their growth instead of oxygen. The growth of SRB in these stagnant conditions leads to the production of hydrogen sulphide and to some bacterial products (often manifested as a lowering of pH). Under some conditions, both oxygen and H_2S can be oxidised to sulphuric acid.

There are, therefore, two main problems resulting from bacterial growth in the offshore structures: hydrogen sulphide production, which is serious to personnel safety; and the production of bacterial metabolites which could give rise to accelerated materials deterioration.

The possibility of ameliorating the effects of bacterial growth within the platform will be discussed.

On the outside of the platform, corrosion processes will also be influenced by biological activity, whether from marine fouling on the platform or from bacterial activity in the sediments surrounding the platform. There are, as yet, few hard data that can be used by the offshore structural designer.

Outside the platform, the real activity of both bacteria and marine growth will also be influenced by platform operational discharges.

Knowledge of bacterial activity and the conditons within the microenvironment created by marine growth is increasing. The next step will be to predict and determine the rate and how this corrosion occurs in practice.

INTRODUCTION

Biological problems occurring with offshore oil operations are often associated with damage to the environment. However, the biological reactions may also affect offshore operations. These biological reactions will take place both inside and outside the structure.

INSIDE THE INSTALLATION

Experience offshore has shown that significant concentrations of hydrogen sulphide are generated by sulphate-reducing bacteria (SRB) in stagnant water in offshore installations, in particular in those platforms with oil/water storage systems.

The formation of stagnant conditions (Fig. 1.4, 1) results from the growth of aerobic bacteria in large volumes of water, in, for example, either the water-filled legs or storage cells of offshore structures. Once the oxygen is consumed, conditions are ideal for the growth of SRB which can use the sulphate present in sea-water for their growth instead of oxygen. The growth of SRB in these stagnant conditions leads to the production of hydrogen sulphide and to some bacterial products (often manifested as a lowering of pH).

Measurements have shown that once stagnant conditions occur, hydrogen sulphide can be generated extremely quickly with the concentration doubling approximately every 10 to 15 h and reaching a peak often at 50/60 p.p.m. in water.

Under some conditions, both oxygen and H_2S can be oxidised to sulphuric acid (Fig. 1.4, 1(c)).

There are, therefore, two main problems resulting from bacterial growth in the offshore structures: hydrogen sulphide production, which is serious to personnel safety, and the production of bacterial metabolites which could give rise to accelerated materials deterioration.

A number of alternative methods have been evaluated to control SRB activity (Table 1.4, 1). The use of a biocide (a chemical used to poison bacteria)

Table 1.4,1

Methods for the control of SRB growth and H_2S production

Legs		
	Chemical addition (biocide)	
	Chlorination	
	Leave legs dry	
	Biological control	
	Use oxygen –	Aerate
		Water replacement
		Drill holes
Storage cells		
	Recirculate sea-water	
	Recirculate sea-water (with chemical addition)	
	Change mode of operation	

a) AEROBIC GROWTH

BACTERIA

O_2 + ORGANIC CARBON ········> CO_2

PRODUCTS

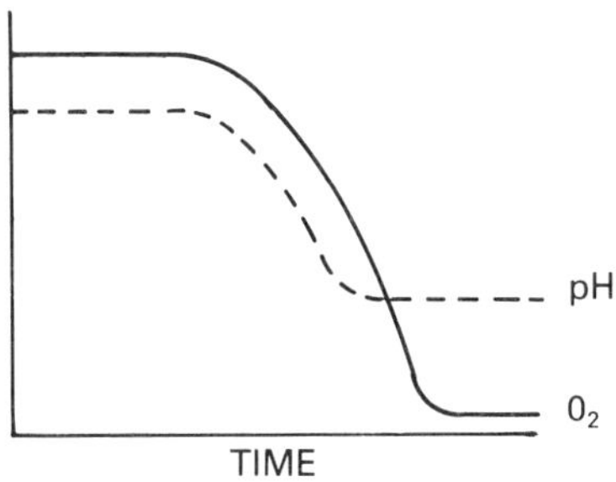

b) ANAEROBIC GROWTH

SRB

SO_4 + ORGANIC ························> H_2S

PRODUCTS

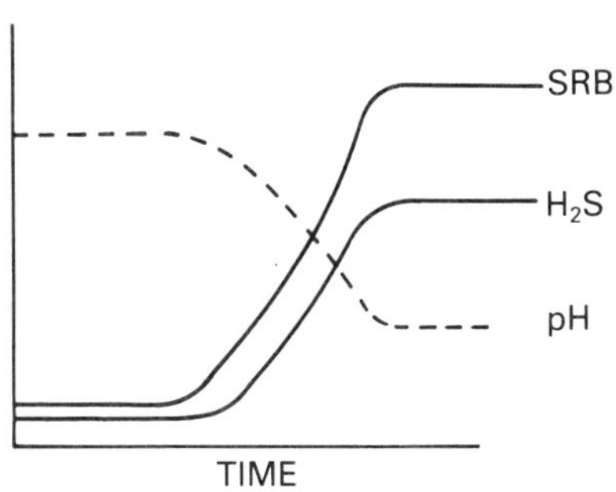

c) SULPHIDE OXIDATION

SULPHUR OXIDISING BACTERIA

H_2S + OXYGEN ························> H_2SO_4

PRODUCTS

Fig. 1.4,1 – Biological mechanisms: (a) aerobic growth, (b) anaerobic growth, (c) sulphide oxidation.

was examined. Whereas biocides can be useful in certain situations, for the large-scale flow-through systems in use offshore in the concrete platforms, there are certain problems:

(a) The effectiveness of conventional biocides for SRB control is doubtful in many situations.
(b) It is difficult to mix biocides adequately with stagnant water.
(c) The use of biocides offshore poses a variety of materials handling problems.
(d) Biocides are costly to use bearing in mind the volumes of water to be treated.
(e) There are severe environmental constraints as the use of biocides could lead to the discharge of unacceptable quantities of poisonous material and a method of control would have to be found which would prevent possibly worse damage to the environment.

Chlorination of water was also considered. However, as chlorine degrades quickly it is not suitable for our systems, and again, could present environmental problems when mixed with soluble hydrocarbons.

Aeration (Fig. 1.4, 2) was found to present an attractive alternative for the control of SRB activity, as dissolved oxygen inhibits SRB growth without associated environmental problems. Maintaining an adequate dissolved oxygen level can be achieved either by directly aerating the water or by changing the mode of operation of the platform to ensure a continuous exchange of fresh sea-water. Systems of this nature have been operating successfully in specific areas on our platforms, for example, in the legs, for over 3½ years.

It has not, however, been possible to maintain control of SRB inside the storage cells of our offshore structures. This means that H_2S (which will be toxic to marine life) and SRB will be discharged with the displacement water during oil fill operation.

It is also important to assess the effect of control conditions, i.e. high dissolved oxygen levels, on the structure. Design allowances have been made for the extra corrosion caused by maintaining a positive dissolved oxygen level and this is being monitored to ensure that it is adequate. The effects of producing micro-environments in sediments at the bottom of the platform and in other stagnant areas where oxygen and H_2S are together, may also lead to the growth of *Thiobacilli* spp. and increased acid production. Reactions of this nature must be recognised and controlled to prevent unacceptable materials attack.

There is also evidence that the rates of concrete and steel erosion may be increased as a result of SRB growth (e.g. by production of an acid environment). At present, we are continuing offshore monitoring work to determine the extent of SRB activity and to establish its true significance offshore as well as carrying out modifications to control SRB activity on offshore structures.

Long-term caretaker projects have also been initiated within Shell Research to examine the microbiology of SRB activity and to determine to what extent the products produced by this activity can cause deterioration of structural materials offshore.

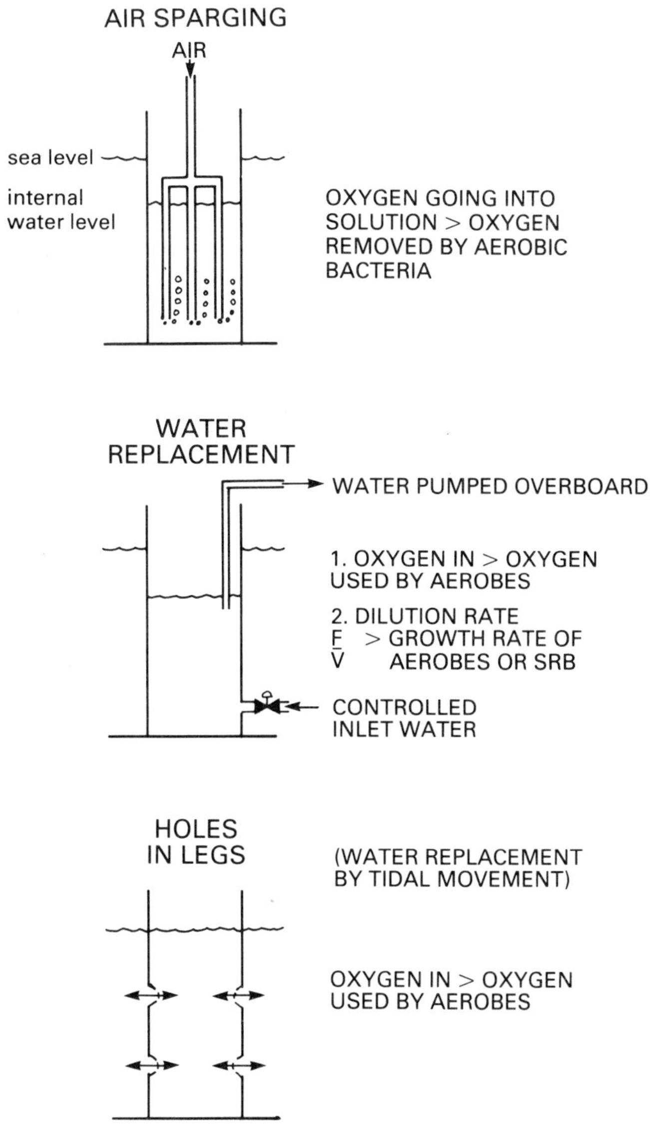

Fig. 1.4, 2 – Methods for maintaining a positive dissolved oxygen level.

OUTSIDE THE INSTALLATION

On the outside of the platform, corrosion processes will also be influenced by biological activity, whether from marine fouling on the platform with associated microbial activity or directly from bacterial activity in the sediments surrounding the platform.

Outside the platform, the real activity of both bacteria and marine growth will also be influenced by platform operational discharges.

DISCHARGES

Major discharges occur from production water, from chemicals used with associated production processes and from normal domestic waste such as sewage.

The major discharge is production water (water that is produced naturally in the reservoir together with oil and gas) which is low in volume at the beginning of the operation, but gradually increases with the breakthrough both of water naturally present in the reservoir and of injected water (i.e. water used to push the oil from the reservoir). This water is cleaned either with conventional gravity separation equipment, such as corrugated plate interceptors or with dissolved gas flotation equipment. This equipment removes the bulk of the insoluble material, but some dissolved oil components may be discharged.

Drilling operations also provide a local pollution source. During drilling, mud is used to lubricate the drill-bit, bring cuttings to the surface and control the reservoir pressure. Dumping cleaned drilling cuttings will cause local effects on the sea-bed (due to settling of the cuttings) and in the water column directly over the sea-bed (due to leaching of the oil from the mud).

Analyses of grab samples taken from the sea-bed (Table 1.4, 2) have shown that this dumping causes considerable local bacterial growth which leads to anerobic sediments and sulphide production by sulphate-reducing bacteria (SRB).

Table 1.4, 2

Distribution of cuttings around Cormorant Alpha – June 1981

Distance from platform leg (m)	75	150	200	250	450
Depth of cuttings (cm)	15-20	0.5	Trace	0	0
Oil content of sediment (p.p.m.)	3500-4000	600-800	150-200	<50‡	<50‡
Depth of sulphide layer† (cm)	20	10-15	Streaky	0	0
Sulphate reducing bacteria (no./gramme)	10^4	10^4	10^4	10^4	10^1

† Sulphide concentration: 30-50 mg/litre.
‡50 p.p.m. level detection.

The depth of the anaerobic regions is essentially the same, whether the sediments are covered by 20 or 0.5 cm of cuttings contaminated with mud. This type of anaerobic bacterial activity will occur regardless of whether oil-based or carbohydrate-based muds are used.

Discharges from production water and the dumping of drilling cuttings will therefore have two major effects. First, in providing nutrients for bacterial growth (as demonstrated by the increased SRB activity in the sediments near the platform) and secondly, in their direct toxic effect on marine life in the immediate locality of the platform which could certainly affect the rate of growth of macrofouling organisms.

Considerable work has been carried out by a number of individuals into both enumerating macrofouling species on offshore installations and into developing techniques to study bacterial growth under the macrofouling and this area is described in the previous two papers of this volume.

CONCLUSIONS

Biological mechanisms can lead to increased materials deterioration both inside and outside offshore installations. At present biologists have been able to enumerate and study the biological reactions that are likely to take place and are gaining a better understanding of the rate at which those reactions will occur. The next step will be for the engineer to use these data to predict the effect of this activity on the offshore installations, and to follow up with appropriate monitoring studies to determine whether the predicted rate is occurring in practice. The results of such studies will also have implications on materials choice for future offshore structures.

1.5 INFLUENCES OF MICROALGAE ON CORROSION OF STRUCTURAL STEEL

L. A. TERRY and **R. G. J. EDYVEAN**†, Offshore Marine Studies Unit, Department of Zoology, University of Aberdeen, Aberdeen, Scotland

INTRODUCTION

The implications of marine fouling growth on North Sea oil and gas platforms have only recently been investigated. The nature of macrofouling communities developing on such structures has already been described by Ralph and Troake [1.5,1], Goodman and Ralph [1.5,2], Goldie [1.5,3], Kingsbury [1.5,4] and by Forteath *et al.* [1.5,5], [1.5,6].

While the colonisation of an installation varies in its detail (especially for macrofouling) depending on the geographical position and time of year, the initial fouling sequence tends to follow a set pattern. Dissolved organic material, which is initially adsorbed on to surfaces [1.5,7], [1.5,8], may be an important factor in providing nutrients, encouraging the initial settlement of bacteria and other micro-organisms. Bacteria can be detected on a surface after only a few hours immersion and these will quickly form dense colonies with copious mucilage [1.5,9]. Hendey [1.5,10] reported 30 million bacteria per gramme of mucilage.

Diatoms are the first algae to settle on the substratum and are numerically second in importance to bacteria. These are closely followed by blue-green algae, protozoa and the spores of larger algae. This microfouling assemblage makes up the 'slime film' widely reported on ships' hulls and other submerged structures. The copious mucilage produced by the bacteria and algae of this film attracts large amounts of detritus and its inherent adhesive nature may encourage the settlement of other organisms. Mixed bacterial and diatomaceous slimes up to 2 mm thick and containing 137 000 diatoms per sq. cm have been reported [1.5,10].

Edyvean and Terry [1.5,11] recorded 270 000 diatoms per sq. cm in 28 days on substrata immersed in Cowes harbour, Isle of Wight and up to 70 000 per sq. cm on substrata immersed along the coast of Northumberland. These figures are for newly immersed substrata; partially cleaned surfaces show more rapid and dense colonisation due to invasion from surrounding areas. Little difference in the numbers of species of microalgae settling on a variety of different substrata were found and it is probable that mucilage plays an important role in the masking of any differences between substrata. This mucilage and trapped material builds up and may form dense layers over a substratum. Goldie [1.5,3]

† Present address: Department of Plant Biology, University of Newcastle upon Tyne, Newcastle upon Tyne, NE1 7RU, UK.

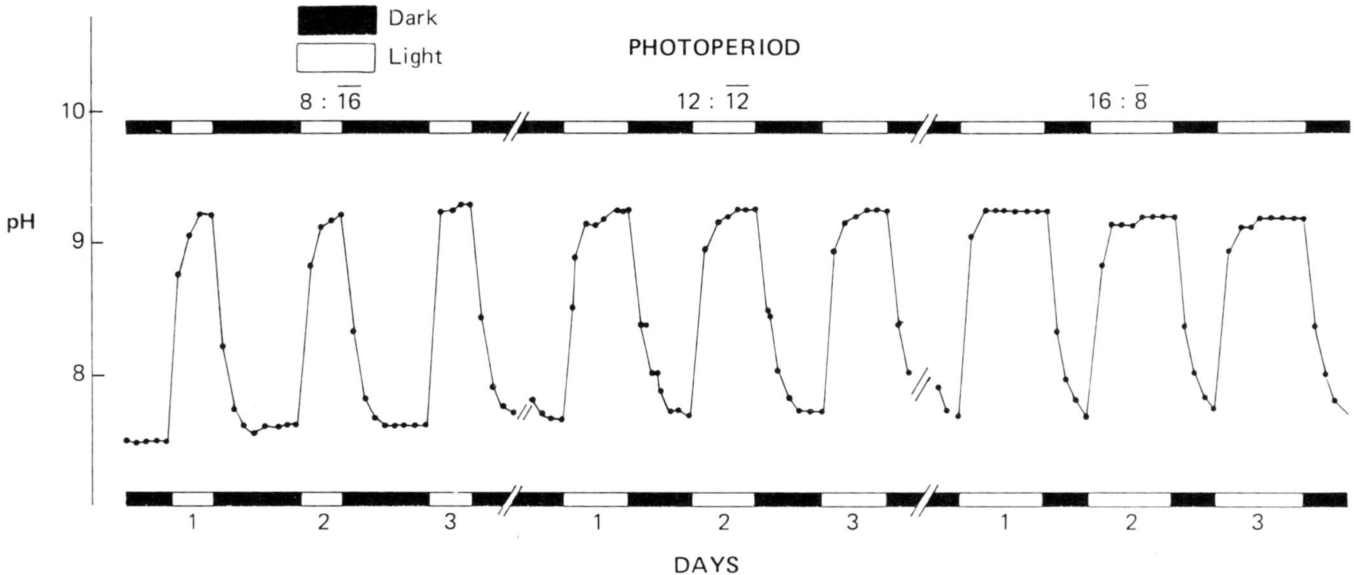

Fig. 1.5,1 – (A) Diatom settlement on paint. 100 days immersion (× 2000). (B) Diatom and blue-green algae on unprotected steel. 100 days immersion (× 2000). (C) Diatoms and blue-green algae on protected steel. 100 days immersion (× 2000). (D) Holdfast of a young *Enteromorpha* plant coming away from a protected steel substratum, removing the surrounding scale. (× 200. H = holdfast. Sc = scale. S = steel.)

has reported a turf-like layer up to 4 cm deep of detrital material, mussel byssus threads, etc. cemented by diatoms, other microalgae and associated mucilage, on a platform in the North Sea. Figure 1.5,1 shows scanning electron micrographs of the settlement of microalgae, associated mucilage and detritus on coal tar epoxy paint, and on unprotected and protected steel after 100 days immersion.

THE INFLUENCE OF MICROALGAL FOULING ON PROTECTIVE COATINGS

The corrosion of steel structures in the North Sea is discouraged by the use of cathodic protection methods (impressed current or sacrificial anodes) and paint coverings. Algal rhizoids, even at a very early stage, have been shown to penetrate certain anti-fouling paint films [1.5,12] and any cracks or breaks in the surface will be taken advantage of by micro- and macroalgae for secure attachment. No paint films are completely impervious and aggressive sulphides have been shown to pass through paint films [1.5,13]. Any breaks in the paint surface will be taken advantage of as access points to the anaerobic environments underneath by potentially aggressive bacteria.

The calcareous scale, deposited on the steel by the action of the cathodic protection systems, may be considerably disrupted by inclusion of settled micro-organisms, forming a packet of potentially aggressive organic material as the scale forms round such organisms. (Figure 1.5, 2 shows a diatom trapped in the scale.) The removal of attached larger algae and invertebrate fouling can break away considerable amounts of the scale [1.5,11]. Figure 1.5,1(d) shows the holdfast of a young *Enteromorpha* plant coming away from a protected steel surface and removing some of the scale. Disruption of this kind, possibly caused by storm action or cleaning could cause problems if the cathodic protection system becomes inoperative for any reason. The exposed area will form an anodic area to the larger area still protected by the scale and severe localised corrosion may result.

MEASUREMENT OF pH OF THE ALGAL/SUBSTRATUM INTERFACE

Algae, even microalgae, modify the pH of the sea-water surrounding them as a result of photosynthesis during the day and respiration at night. This would normally have little effect on a substratum as the frondose nature of macroalgae mean that this activity is taking place in flowing sea-water which rapidly buffers any changes. However, considerable changes are detected in the trapped water around young algae or under ‘mats’ of microalgae such as diatoms or blue-green algae. These microalgae produce copious mucilage and thus trap a discrete body of water at the surface of the substratum.

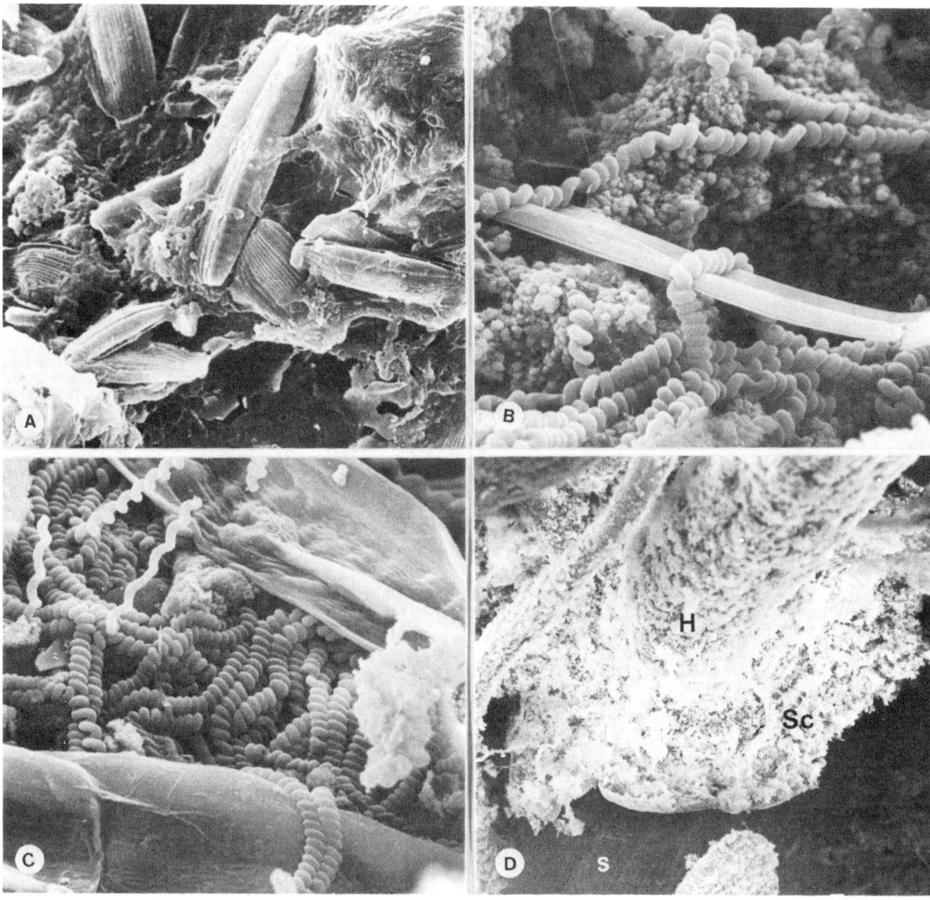

Fig. 1.5, 2 – Diatom trapped in the scale precipitated on a cathodically protected steel surface (× 5000).

pH values at the algal/substratum interface have been measured for various algal cultures [1.5,11] [1.5,14]. Figure 1.5, 3 shows a typical plot of pH changes between day and night under a mat of the blue-green algae *Oscillatoria* for various light/dark regimes. A maximum pH is reached during the light period and this tends to be held, regardless of its length. This maximum pH is a function of the metabolism of the algae and is different in each species: for *Oscillatoria* it is between 9.4 and 9.6 [1.5,11]. For some diatoms the pH maximum is between 9.6 and 10 while *Enteromorpha* cultures can produce a maximum pH above pH 10 [1.5,14]. The pH low achieved during dark periods rarely goes below that of the sea-water (7.8-8.2), though the normal buffering capacity of trapped sea-water can be broken down and then pH fluctuations will be at their greatest

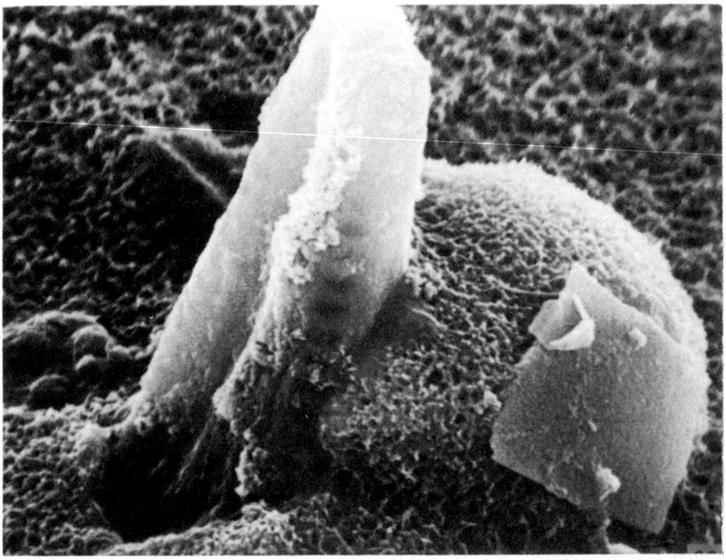

Fig. 1.5,3 – pH changes under an established mat of the blue-green alga *Oscillatoria* sp. (Irradiance 6 W^{-2}, 15 °C.)

[1.5,14]. The pH under decomposing algal cultures has been measured [1.5,11] and pH values as low as pH 1.8 have been recorded under decomposing diatom cultures which could have a considerable effect on the underlying steel.

While high pH values may have a protective effect in encouraging the deposition of calcium carbonate, slightly lower pH values may be aggressive as the bicarbonate then present will react with ferrous ions to form soluble compounds $[Fe(CO_3)_2]^{2-}$ which can greatly accentuate the corrosion of the metal causing pitting [1.5,15].

POLARISATION MEASUREMENTS

Direct measurement of the influence of an environment on steel can be made by the use of potentiostatic polarisation methods. The current response resulting from changes in applied potential to an electrode is plotted and a prediction of the corrosivity of the environment may be made from the study of the polarisation curves produced [1.5,16], [1.5,17]. The reactions at the metal surface will be influenced by its surface characteristics and the electrolyte, which in turn may be influenced by the presence of fouling organisms present in the medium or growing on the surface.

These methods have been used to monitor a variety of microalgae and young stages of macroalgae. (Edyvean and Terry, in press.) It is known that sulphate-reducing bacteria can be found associated with healthy *Enteromorpha* plants [1.5,18]. No evidence of their action was found in young cultures. Under anaerobic conditions where the plants were decomposing, their presence was indicated by the evolution of hydrogen sulphide and the presence of ferrous sulphide and the corrosion reactions were enhanced by cathodic depolarisation. Healthy cultures of the blue-green algae *Oscillatoria* were found to enhance corrosion by anodic depolarisation, possibly by a similar method to that reported for aerobic bacteria under microanaerobic conditions, i.e. the breakdown of protective iron oxide films [1.5,19].

CONCLUSIONS

The microalgal community, producing copious mucilage, may have beneficial effects in consolidating the underlying scale or corrosion product and restricting the access of oxygen. However, the local chemistry will be complex and will vary with secretions, decay products and associated bacterial action. In particular such action will result in the production of sulphide compounds and these could cause serious localised corrosion. The mucilage produced by the algae and bacteria will serve to concentrate these products and localise the affected areas, causing discontinuities giving rise to galvanic couples and localised corrosion.

ACKNOWLEDGEMENTS

This work was carried out at the University of Newcastle upon Tyne and financed by a research grant awarded to Dr B. L. Moss by the Science and Engineering Research Council, Marine Technology Directorate. R.G.J.E. was supported by an S.E.R.C. CASE award sponsored by British Petroleum. The authors wish to acknowledge the assistance of British Petroleum in providing some of the materials used and in carrying out the X-ray crystallographic analysis.

REFERENCES

[1.5,1] Ralph, R. and Troake, R. P. Marine growth on North Sea oil and gas platforms. Paper OTC 3860. *Proceedings of the 12th Annual Offshore Technology Conference*, Houston, Texas. 1980, **4**, 49–52.

[1.5,2] Goodman, K. S. and Ralph, R. Animal fouling on the Forties platform. In *Marine Fouling of Offshore Structures.* Society for Underwater Technology, London, **I**, 1981.

[1.5,3] Goldie, B. P. F. Assessment of marine fouling on gas platforms "WE". In *Marine Fouling of Offshore Structures.* Society for Underwater Technology, London, **II**, 1981.

[1.5,4] Kingsbury, R. W. S. M., Marine fouling of North Sea installations. In *Marine Fouling of Offshore Structures*. Society for Underwater Technology, London, **I**, 1981.

[1.5,5] Forteath, G. N. R., Picken, G. B., Ralph, R. and Williams, J. *Mar. Ecol. Prog. Ser.*, 1982, **8**, 61-68.

[1.5,6] Forteath, G. N. R., Picken, G. B. and Ralph, R. Patterns of macrofouling on steel platforms in the central and northern North Sea. In this volume, p. 000.

[1.5,7] Loeb, G. I. and Niehof, R. A. *J. Mar. Res.*, 1977, **35**, 283-291.

[1.5,8] Baier, R. E. Substrate influences on the adhesion of micro-organisms and their resultant new surface properties. In *Absorption of Micro-organisms to Surfaces*. Edited by Brittin, G. and Marshall, K. C. John Wiley, New York, 1980, 59-104.

[1.5,9] Dempsey, M. J. *Botanica Marina*, 1981, **24**, 185-191.

[1.5,10] Hendey, N. I. *J. Roy. Microscop. Soc.*, 1951, **71**, 1-86.

[1.5,11] Edyvean, R. G. J. and Terry, L. A. The influence of microalgae on the corrosion of structural steel used in the North Sea. In *Biodeterioration 5*. Edited by Oxley, T. A. and Barry, S. John Wiley, 1982.

[1.5,12] Moss, B. L. Observations on the breakdown of paint surfaces by ship-fouling algae. In *Proceedings of the Third International Congress on Marine Corrosion and Fouling*. Edited by Acker, R. F. *et al.*, National Bureau of Standards, Gaithersburg, Maryland, USA. 1974, 33-38.

[1.5,13] Williams, M. E., King, R. A. and Miller, J. D. A. *J. Oil Colour Chem. Assn.*, 1973, **56**, 363-368.

[1.5,14] Terry, L. A. and Edyvean, R.G. J. *Botanica Marina*, 1981, **24**, 177-183.

[1.5,15] Davis, D. H. and Burstein, G.T. *Corrosion*, 1980, **36**, 416-422.

[1.5,16] Booth, G. H., Robb, J. A. and Wakerley, D. S. The influence of ferrous iron on the anaerobic corrosion of mild steel by actively growing cultures of sulphate-reducing bacteria. In *Proceedings of the Third International Congress on Metallic Corrosion*. Moscow, 1966. Published 'Mir', Moscow, 1969, **II**, 542-554.

[1.5,17] Salvarezza, R. C. and Videla, H. A. *Corrosion*, 1980, **36**, 550-555.

[1.5,18] Baas Becking, C. M. G. and Wood, E. J. F. Biological processes in the estuarine environment. 1, 11 *Konoklijke Nederlandse Acadamie van Wetenschappen*. Proc. Ser. B., 1955, **58**, 168-181.

[1.5,19] Obuekwe, C. O., Westlake, D. W. S., Plambeck, J. A. and Cook, F. D. *Corrosion*, 1981, **37**, 461-467.

1.6 FOULING AND CORROSION OFF THE OUTER HEBRIDES

M. J. PICKEN and C. GRIER, Scottish Marine Biological Association,
Dunstaffnage Marine Research Laboratory, Oban, Scotland

ABSTRACT

The problems associated with marine fouling and corrosion are common to many types of underwater structures. The National Wave Energy Programme would involve the deployment of devices in the north-east Altantic west of the Outer Hebrides and studies have been concentrated into calculating the likely effects of the marine environment on these structures. Among the effects of marine fouling, the modification of corrosion behaviour of affected steel is considered to be one of the most important and least understood. Some preliminary experiments have been conducted to study the effects of the Hebridean environment on corrosion of mild steel the results of which should provide additional technical guidance data if hydrocarbon exploration should ever take place in that area.

Wave energy is one of the several possible alternative energy sources which has been examined intensively over the past five years. Along with several other sites around the UK coastline, the waters west of the Outer Hebrides were selected for detailed investigation into the overall suitability of the area for a full-scale device demonstration. These studies were funded by the Department of Energy via the Energy Technology Support Unit at Harwell which also directed the present studies on fouling and corrosion.

Data was obtained from panels deployed on moorings 10 km west of South Uist from 1978 until the time of writing. Additional data were collected by the examination of navigation bouys belonging to the Northern Lighthouse Board in the sea area between the Inner and Outer Hebrides (Fig. 1.6,1). Although these particular bouys were not in the wave energy area they provided very useful collateral information on the biology of the main fouling organisms over some wider ranging environments. The results of the data analyses were then applied via varying data confidence ratings to actual fouling predictions on specific wave energy devices. However, the considerable difficulties of extrapolating limited field observations into actual longer term predictions had to be considered in light of North Sea experience.

The Hebridean wave energy area differs from the oil and gas fields of the North Sea in a number of ways:

(i) The area is affected by a north-easterly flowing residual water current reinforced most of the year by wave-induced orbital currents.

(ii) The area is typically shallow, <100 m deep, and lies within 20 km of the Outer Hebridean coastline; infection from parent stock is therefore high.

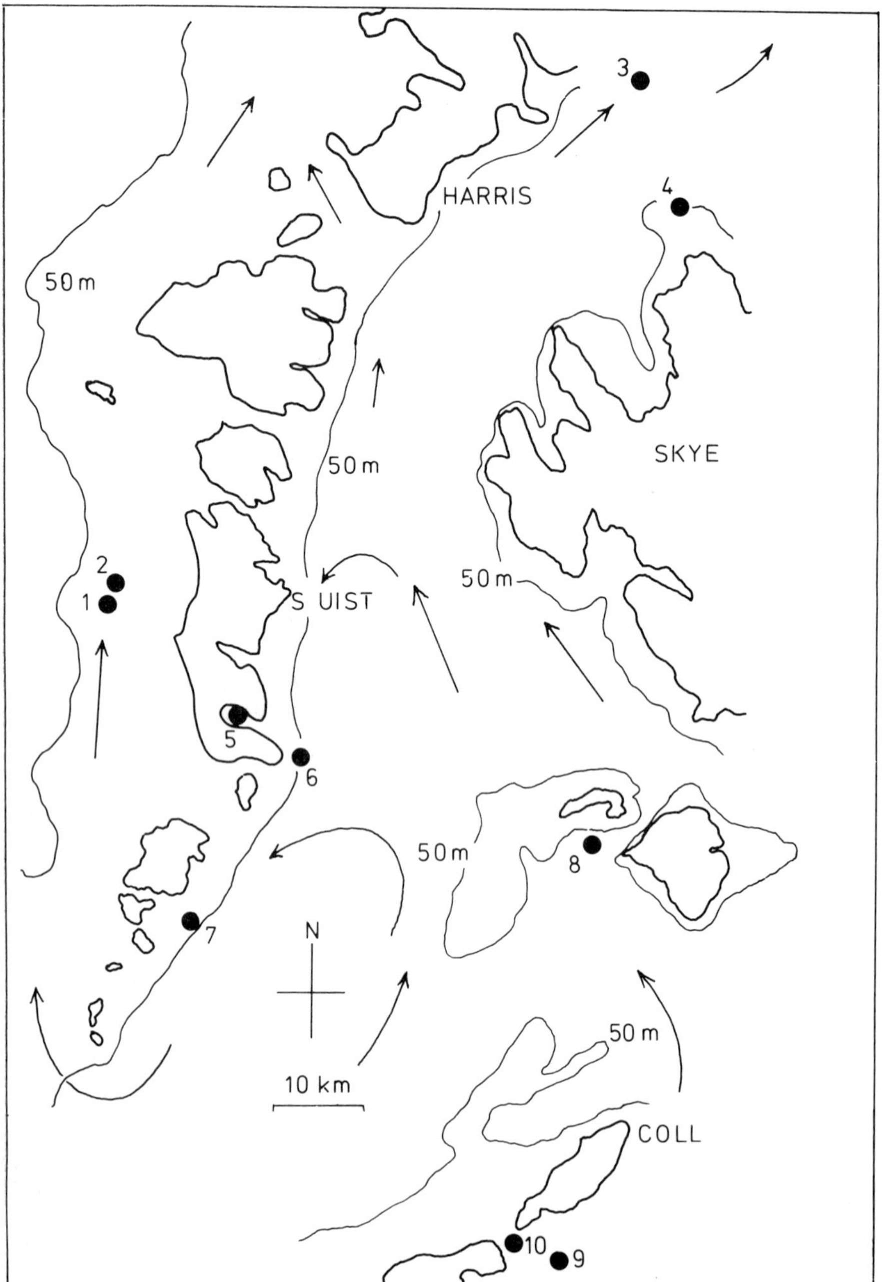

Fig. 1.6,1 – The survey area to show stations worked, the 50 m isobath and summer surface currents.

(iii) The sea-bed over much of the area is geologically continuous with the rocks of the Outer Hebrides and is represented by a very gently shelving plateau of Lewisian gneiss, much dissected and corrugated by the effects of previous glaciations.
(iv) High optical clarity of the water enables seaweeds, in particular the kelp *Laminaria hyperborea*, to extend to a depth of 30 m.

Over 100 sessile fouling species were identified out of which about fifteen occurred commonly at most of the sampling sites. Many of the species identified are also found in fouling communities in the North Sea and reinforce the general observation that there are marked similarities in flora and fauna on man-made objects in coastal waters of the UK.

The results have been described in detail by Picken *et al.* [1.6,1] and may be summarised thus:

(i) Kelp (*Laminaria hyperborea*), mussels (*Mytilus edulis*), hydroids (*Tubularia larynx*) and other soft fouling such as anemones and sea-squirts formed the climax fouling constituents after three years.
(ii) Barnacles (mainly *Balanus crenatus*) formed heavy settlements during the first and second years after deployment but were considered to be ephemeral in nature and of little importance once mussel communities had been firmly established.
(iii) Vertical distribution of fouling on arrays and moorings was kelp (0-30 m but the general lower limit occurring at about 20 m), mussels (0-30 m) and hydroids (20 m downwards).
(iv) Even in early years isolation from infection appears to be minimal due to ease of dissemination by prevailing water currents.
(v) Once initially established the main fouling groups exhibited growth rates similar to those groups in the North Sea. (Mr A. Pipe, personal communication.)

The most important difference when compared to the North Sea was the early occurrence of the kelp *Laminaria hyperborea* on the moorings west of the Outer Hebrides plus the common occurrence of some southern species such as the stalked barnacle (*Lepas anatifera*) and certain calcareous bryozoans. Although the observations were from one location west of South Uist there is a very generalised pattern in the distribution of this kelp along the entire Outer Hebridean west coast, the sea-bed being dominated by a dense kelp monoculture extending to the 30 m isobath about 4 km off-shore. It is thought that kelp populations can be much extended in this area by the piecemeal distant transport of whole plants during the autumn prior to spore-release later in the winter. Although net transport is likely to be mainly in a north-easterly direction, occasional strong easterly winds could extend the infection potential westwards.

Underwater stereo-photogrammetry has been successfully used as a tool in the non-destructive investigation of rocky benthic communities in Scandinavia by T. Lündalv and others. This method is particularly well suited to the monitoring of biological change on man-made structures and a system has been developed at Oban which can be used to monitor short and long term biological change in a number of situations. Among those, the system has been applied over the past year to the weekly investigation of the build-up of macrofouling on cathodically protected plates at a sheltered sea-loch site. Analysis of some early results has demonstrated the instrument's future application in the monitoring of surface structure change and fouling composition in planned future corrosion experiments. A technical description of the system is given in the final section.

CORROSION STUDIES

The moorings for the lightweight buoyed arrays used in the initial fouling experiments were based on the design of the SMBA continental shelf moorings which have been successfully used in hydrographic studies in the north-east Atlantic over the past eight years [1.6, 2]. However, none of these moorings was subjected to the physical extremes experienced by the fouling arrays which were placed in shallower and more turbulent water.

Examination of a mooring system which had failed indicated that excessive corrosion rather than abrasion had been instrumental in the failure. This incident stimulated the present corrosion studies which are taking place in that particular sea area. These studies started in 1980 and it must be stressed that the results obtained are of a very tentative nature and serve to outline some particular problems which will require much more detailed investigation in the future.

Details of the SMBA† experiments have been described by Bowers [1.6, 3] and Picken *et al.* [1.6, 1] but a realistic interpretation of the results will only be meaningful once the present and proposed future experimental results have been analysed.

The design values recommended for the protection of cold bare steel in the North Sea are of the order of 100–150 mA/m^2. However, it is assumed that this is an average figure and it might be reasonably expected that, in the absence of a passivating coating, the current demand would be higher near to the water surface where oxygen flux over cathodic areas is higher. In the deeper, less turbulent and possibly less oxygenated water one could expect the current demand to be reduced. Figures of >200 mA/m^2 have been stated for protection in the Northern Continental Shelf area where even colder and more turbulent conditions could be expected.

The basic framework of the first results can be tentatively outlined. Three

† Scottish Marine Biological Association.

experiments were deployed. The first was composed of mild-steel plates with zinc anodes connected via 2.2, 8.2 and 39 Ω resistances respectively plus an unprotected control. Cathode to anode surface area ratios were 11 : 1. The panels were mounted at a depth of 1.5 m 10 km west of South Uist and examined after one year. Based on anode weight loss and examination of cathodes it was estimated that an average current density of $>$300 mA/m^2 failed to provide protection [1.6,4].

The second experiment consisted of a set of identical panels which were deployed at a depth of 5 m in Loch Creran which is a sheltered mainland sea-loch. Weekly assessment of the development of corrosion and fouling was monitored *in situ* by underwater photogrammetry. Although the experiment is still continuing results after 18 months show complete protection on the cathodically protected plates, negligible anode consumption and a rich and dense fouling community. As the anode loss has been minimal the average current density is likely to be very low. These first observations are consistent with the low water velocities ($<$10 cm/sec) and high levels of fouling.

The third experiment involved the use of a modified Anderaa RCM 5 current meter which measured water speed, temperature and the corrosion current and corrosion potential of a mild-steel plate protected by a zinc anode and incorporating a second reference anode made of similar anode quality zinc [1.6, 4], [1.6, 3]. The experiment was deployed at a depth of 24 m close to the first mooring. The two sequential experiments showed the normal behaviour of corrosion current and potential with time, the former asymptoting to about 100 mA/m^2 which appeared to have provided almost complete protection in spite of an absence of calcareous scale. However, the January to April record showed, in addition, many occasions when the corrosion current was in phase with water speed but out of phase with the potential. This occurred only on the first record (January to April) and not on the second (April to July). The first record shows low winter/spring temperatures and generally higher water speeds and the anomalous current and potential behaviour appears to be associated in some way with the higher water speeds and corresponding high level of oxygen flux. Detailed critical study of the potential record eliminated any suggestion of reference electrode drift.

Preliminary discussions have suggested that the unusual electrode behaviour could be interpreted in terms of cathodic depolarisation associated with some manifestation of very high oxygen flux over the electrode surfaces. Without attempting to enter into a premature presumptive analysis of early information it may be useful to look at the wave and water velocity picture of the two depths representing failure of protection at 300 mA/m^2 and near protection at 100 mA/m^2.

In that sea area the tidal streams are weak, $<$40 cm/sec, but wave-induced orbital currents greatly exceed this figure most of the year near the surface and during much of the autumn and winter at depths of 24 m. If recorded water speed at 24 m and significant wave heights are plotted out, a straight line can be drawn with the water speed intercept corresponding to the residual water speed which is about 12 cm/sec.

From analysis of wave records for that area it is seen that the significant wave height is >1 m during about 90 per cent of the year. This wave height has an associated surface orbital speed of 40 cm/sec for an 8 sec period wave and 65 cm/sec for a 10 sec wave. Similarly from the data, significant wave heights of >4 m can be expected to induce orbital speeds of >45 cm/sec at depths of 24 m but these larger waves prevail for less than 10 per cent of the year. Thus it would seem that if these high water speeds are associated with depolarisation behaviour then they occur over most of the year in the surface water.

These first results leave much unanswered and serve mainly to stimulate further field studies. It is hoped that the following proposed experiments may attempt to answer some of these questions.

(1) A repetition of the current meter experiment over a longer period representing a greater seasonal variation in water temperature and with marine fouling allowed to develop over a longer period.
(2) Examination of some corrosion parameters at much higher water speeds such as those encountered at some sea-loch entrances in an attempt to understand more about the specific requirements for the unusual current and potential behaviour noted earlier.

UNDERWATER PHOTOGRAMMETRY SYSTEM EQUIPMENT DESCRIPTION

A pair of Nikonos IVa amphibious cameras fitted with UW-Nikkor, 15 mm, f 2.8 lenses were mounted with parallel optical axes on a common sub-frame within a main-frame designed for photogrammetric analysis of the resultant stereo image. The main-frame which was box-shaped covered an area of 50 cm × 50 cm and was illuminated by a 'Sub Sea 150' submersible electronic flash unit (Fig. 1.6, 2). The complete assembly was adjusted for neutral buoyancy by the addition of two 150 mm diameter plastic trawl floats and in this state was easy to manoeuvre and operate by free swimming divers.

The application of photogrammetry underwater allows the introduction of a high precision three-dimensional measuring technique without interference of the subject. A permanent record is obtained and any dimension within the area of stereo coverage can be established at a later date.

The Nikonos IVa cameras were operated by means of adjustable pawls mounted on a common stainless steel shaft. A small lever at one end of the shaft could be operated by hand to fire the shutters. A spring returned the level to its 'ready' position. The individual camera films were then wound on separately prior to the next exposure. Adjustment of the pawls allowed the camera shutters

Fig. 1.6, 2 – Complete assembly for underwater photogrammetric work; showing main-frame, sub-frame, cameras and buoyancy units.

to be fired almost simultaneously. Electronic flash synchronisation was achieved by having the left-hand camera shutter set on 'B' (brief) and the right-hand camera on 'M' (manual), the normal position for electronic flash. Fine adjustment of the pawls allowed the left-hand camera shutter to open a few milliseconds before the right-hand camera which was connected to the electronic flash unit. Immediate release of the shutter lever allowed the left-hand camera shutter to close. The few milliseconds of extra exposure allowed to the left-hand camera does not substantially affect the exposure made in the low ambient light level of the undersea environment.

The extreme wide angle of the 15 mm lens (94° across the negative diagonal) is very advantageous in the underwater situation in that it allows the area to be covered from the shortest possible through water distance. The maximum image quality could thus be obtained if the light source was correctly positioned to minimise back scatter from solids in suspension in the water. This was achieved by having the electronic flash mounted in a suitable position displaced from the camera lense by a bracket and angled towards the subject at about 45°.

The 'Sub Sea 150' electronic flash unit was set to full power at its maximum angle and an exposure of f8 for Kodak 'Ektachrome 64' Professional Film rated at 100 ASA allowed adequate depth of field. 'Ektachrome 64' was chosen for its ease of processing. Its resolving power and colour contrast proved to be quite adequate for the current requirement.

The camera sub-frame with the electronic flash bracket was supported and held in position by two columns above the photogrammetry reference or control frame. This frame was box-shaped with internal dimensions 50 cm X 50 cm (covering ¼ sq. m) and 10 cm high. It was constructed of square section aluminium tubing and an extension plate on either side carries a writing tablet, depth gauge, magnetic compass, colour separation chart and thermometer. The area within the frame was covered in stereo by the overlapping photographs which utilised 66 per cent of the available negative area for this. Instrumentation data was recorded monoscopically on the remaining negative areas of one or other of the cameras.

Target points were laid at approximately 5 cm intervals round both the top and bottom levels of the frame and provide control points for three-dimensional analyses of the stereo pair photographs.

When properly calibrated the system is capable of full photogrammetric analysis to a high order of precision. In practice this has not normally been warranted by the requirements which have on the whole been satisfied with examination by stereoscope of the transparencies for species identification and visual estimates of areas covered.

REFERENCES

[1.6, 1] Picken, M. J., Bowers, R. and Shaw, T. L. Marine fouling and corrosion off the Hebrides. *In press.*

[1.6, 2] Edelsten, D. A current meter mooring for use on the Scottish continental shelf. *S.M.B.A. Internal Report*, 1980, No. 34.

[1.6, 3] Bowers, R. Modifications to an Anderaa RCM 4 current meter for the measurement of corrosion current and potential. *S.M.B.A. Internal Report*, 1981, No. 40.

[1.6, 4] Wilkins, N. J. M. Cathodic protection of metals in sea-water. 1982. *In press.*

[1.6, 5] Arup, H., Christensen, C. and Møller, J. Corrosion and cathodic protection in Arctic water. In North Sea Corrosion Conference Transactions. *I. Mar. E.*, 1979, **91**.

CHAPTER 2

Fabric maintenance

2.1 UNDERWATER INSPECTION FOR CORROSION

B. BALMER, Chevron Petroleum (UK) Ltd, Aberdeen, Scotland

ABSTRACT

The primary objectives of underwater inspection are to find corrosion, provide sufficient information to relate to structural integrity and act as a baseline for monitoring. This paper briefly outlines some of the main considerations for corrosion and underwater engineers to enable such objectives to be fulfilled. The discussion applies mainly to offshore steel and concrete oil production platforms operating in the North Sea.

MARINE CORROSION

Corrosion is broadly defined as the deterioration of a material due to a reaction with its environment where deterioration means, or implies, a change in the structural properties. Identifying and quantifying the extent and significance of corrosion concerns corrosion and inspection engineers. Steel and concrete offshore structures exist in an environment of oxygen saturated sea-water, which is particularly detrimental to structural steels. Corrosion will occur if the proper precautions are not taken.

If corrosion of steel in sea-water is considered as an electrochemical process, the three main components of the corrosion cell will be:

(a) Sea-water (electrolyte).
(b) Corrosion site (anode).
(c) Non-corroding site (cathode).

All three components are critical if the corrosion process is to be maintained. If any one component is removed, the process stops. If the chemical nature or structure of a component changes, the process will change. This change may result in an increase or decrease in the corrosion rate.

The existence of the anodes and cathodes in manufactured structural steels is well known. Why and where they exist in steel can be predicted by examining metallurgical, environmental and fabrication factors.

Within a single steel, a number of phases can occur based on iron and carbon solid solutions. These phases have different physical and electrochemical properties which allow the nucleation of many interchanging anode and cathode sites. A number of individual corrosion cells are set up eventually leading to an overall general corrosion visible to the naked eye.

During manufacture mill scale is formed on a steel surface which acts cathodically to steel and can lead to intense surface pitting. Welding of steels can introduce residual stress and cause local physical and composition changes, both factors giving rise to corrosion.

Concentration as well as composition changes can lead to anodes and cathodes being established. Oxygen, salinity and temperature can change with season and depth, leading to long line anode/cathodes. Particular problems exist if oxygen deficiency arises as, for example, beneath marine growth and sediments, in which case bacterial activity may give rise to serious corrosion.

CORROSION PREVENTION

Consider the three-component corrosion cell discussed. Changes to components will result in changes in corrosion rate. Practically speaking, in the North Sea, it is not possible to make significant changes to the electrolyte, although seasonal changes will occur due to natural processes. The main changes can be made to the anodes and cathodes on the steel surface.

Two methods of corrosion prevention are employed in the marine environment and particularly on offshore structures. First, organic surface coatings, such as epoxies, urethanes and neoprenes are applied to the steel. The result is almost total separation of the steel from the sea-water electrolyte. Second, cathodic protection, either by impressed current or sacrificial anode methods may be used. The steel surface in this case is supplied with a surplus of electrons making it more cathodic. The anodic sites become cathodes with respect to the sea-water and corrosion is reduced.

UNDERWATER INSPECTION

Consideration must be given to the design of the structure at an early stage in planning the inspection. This will highlight areas where corrosion is most likely to occur, and where corrosion will have the most significant effect on integrity.

The area of the splash zone, the mud line, welds – particularly those under high cyclic stress – crevices, surfaces operating at elevated temperature, and dissimilar metals in electrical contact will be given particular consideration.

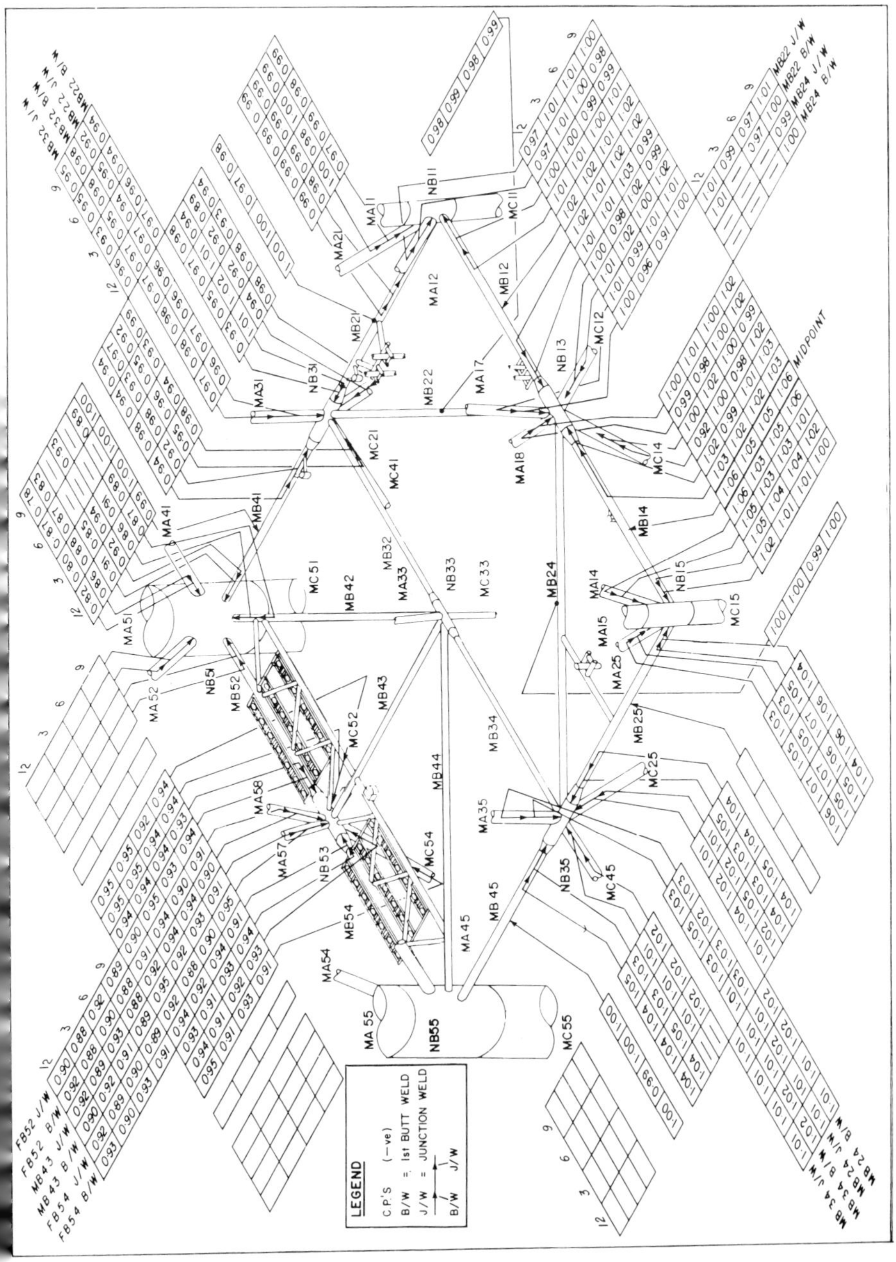

Fig. 2.1,1 – Structure elevation showing three-dimensional reference system (data acquisition).

The economics and practical difficulties of working underwater mean that it is not always possible to inspect every case where corrosion is likely to occur. Access to the work site, type of surface and subsurface support (diver or ROV) and the safety requirements will feature significantly at the planning stage. Principal representative areas for inspection have to be established and sufficient flexibility built into the plan to take account of the results of the original inspection plan.

A summary inspection schedule detailing the scope and procedural requirements of the inspection is written after every consideration has been given to the technical requirements. This schedule will recommend the frequency of inspection, which for most practical purposes will be annual, but may vary depending on the particular demands of the situation.

Underwater inspection requires a controlled approach during implementation due to the physical constraints of the environment. Selection of suitably trained personnel is essential to establish a standard of quality information. Subjective judgements must be avoided, and to this end, and to ensure efficiency, controls are necessary.

First, dimensional controls need to be established. This simply requires some method of referencing a location on a structure. Each component will

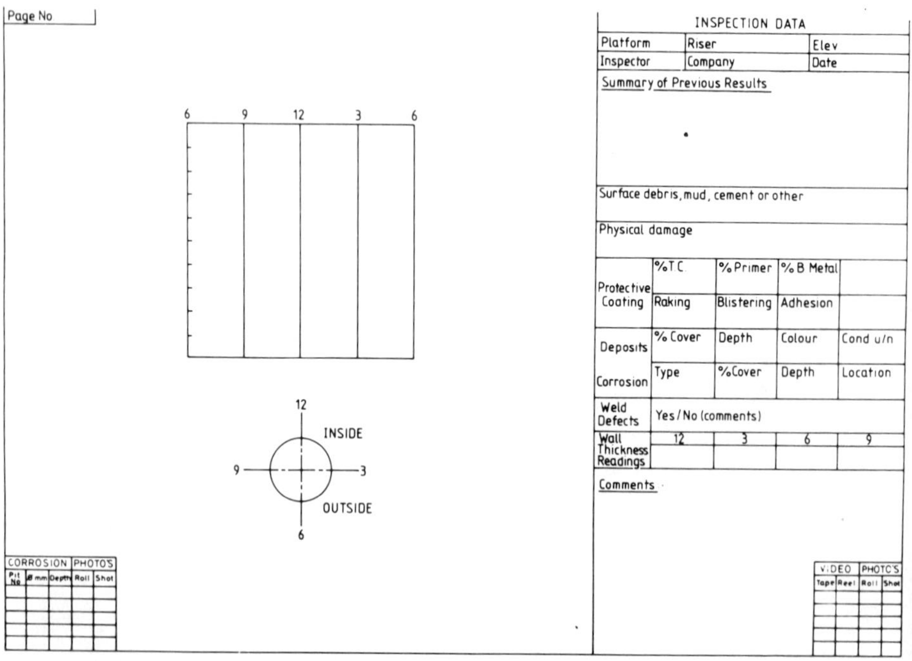

Fig. 2.1, 2 – Detailed data sheet (data acquisition).

have a unique reference code as given on a drawing. An example is shown in Fig. 2.1, 1. On a smaller scale, particular areas such as welds or clamps can be detailed on drawings as shown, for example, in Fig. 2.1, 2. Information can be summarised offshore on such drawings. This is the so-called data acquisition phase.

Second, equipment controls need to be established. Type of cleaning, visual aids such as video and still photography to be employed and physical measurements to be taken are the primary considerations. This is the so-called procedural phase.

EQUIPMENT

It is worthwhile considering the equipment and how it can be used. The correct type of cleaning must be established for the degree of inspection and defect size specified. Before mentioning the types of cleaning techniques employed one must consider what information can be gained about corrosion prior to cleaning. During the cleaning operation, corrosion or calcareous deposits will be cleaned from a steel surface as well as marine growth. It is well known that corrosion products of steel are normally red-brown in colour. It may be very obvious from the colour of the deposits noted prior to cleaning, that corrosion is local to one area. An example is given in Fig. 2.1, 3.† Calcareous deposits may occur as grey or white and generally indicate effective corrosion protection. Sampling and chemical analysis of surface deposits is also a method by which the exact nature of the corrosion process may be established.

The types of cleaning used range from wire brushing to water jetting, or grit/sand entrained water jetting. Silica sand used in the water jet achieves a matt grey surface. This surface finish is ideal for visual and photographic inspection. Figures 2.1, 4, 2.1, 5 and 2.1, 6 show steel surfaces after various types of cleaning have been carried out.

Visual examination is always the first line of inspection. This sets the scene for further inspection, typically providing information regarding general or localised corrosion, shallow or intense pitting, etc.

CCTV (video) facilities provide a monitor for the inspection engineer and in all cases, form the basis of the engineer's visual examination and a permanent record of findings. Both monochrome and colour cameras are now available for underwater use. The use of colour will greatly improve the inspection for corrosion.

Thirty-five mm and 70 mm colour still photography are both used extensively for recording the findings of inspection. Generally speaking, finer details can be observed on still photographs and in some cases by the use of stereo pairs for which photogrammetric analysis may be carried out.

† Figures 2.1,3–2.1,6 are printed from original Ektachrome colour slides taken with a Nikonos 4 camera hand held by diver.

Fig. 2.1, 3 – Surface deposit prior to cleaning. Red/brown deposits locate area of corrosion.

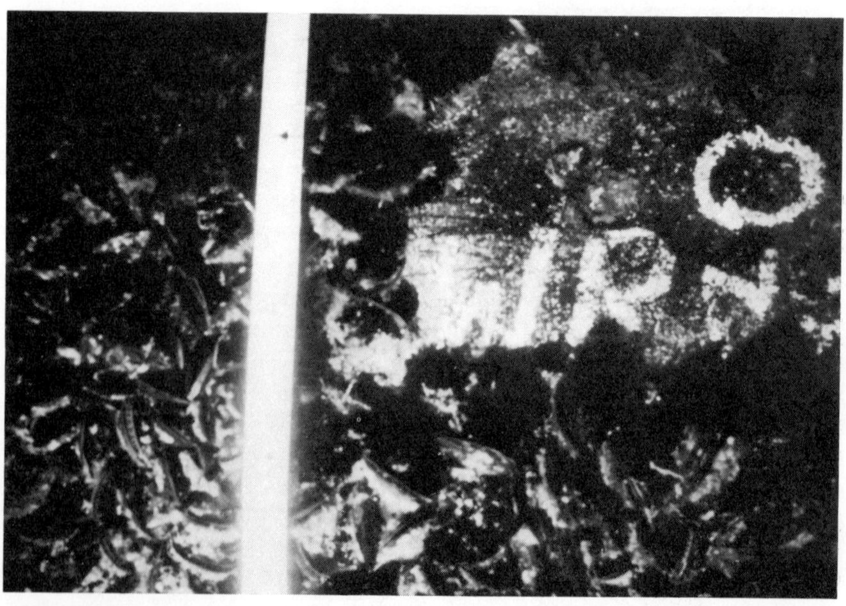

Fig. 2.1, 4 – Steel surface after wire-brush cleaning.

Fig. 2.1, 5 – Steel surface after high pressure water jet cleaning.

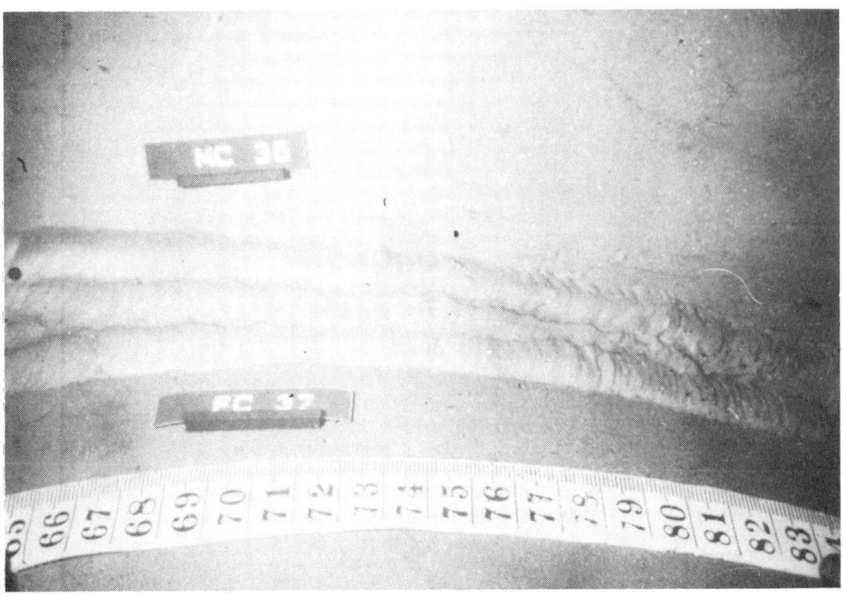

Fig. 2.1, 6 – Steel surface after grit/sand entrained water jet cleaning.

Wall thickness measurement is carried out by ultrasonic techniques using conventional topside equipment either marinised or purpose built for diver use. Accuracy is as for equipment used on the surface, but controls are required to ensure correct use within the constraints of the underwater environment.

Tape measures, pit gauges, profile gauges and various methods of impression moulding are used to measure and monitor corrosion. Recent advances in the application of underwater stereo photography have resulted in the use of photogrammetric analysis to measure corrosion. Accuracy to 0.1 mm may be achieved using this method.

Remote silver/silver chloride reference electrodes (half cells) with surface voltmeter systems are used for the measurement of steel cathode potential. These methods differ mainly in the method of making electrical connection to the steel cathode. In the case of the remote cell, connection is made to the structure above water and locally, by a probe in the diver held system. Both systems are used extensively with satisfactory results.

REPORTING

The report will provide a formal record of findings. Following the record, an analysis of the information or data should be presented showing what relationship exists to the original design or previous information. Certain information or data may require analysis to indicate future needs both in the short and long term.

A great deal of information is generated during underwater inspection. This information represents an investment. The return on this investment can only be realised if the information is readily available at all times in an easily used form. For this reason development of a computer data base is an obvious choice for underwater inspection reporting. This area is receiving an increased amount of attention.

SUMMARY

The main objective of underwater inspection is to locate, quantify and report deterioration. Corrosion is now, and will continue to be, a major cause of deterioration of offshore structures. Finding corrosion underwater is not simply a matter of divers swimming around looking at every part of a structure. Engineers with a knowledge and experience of corrosion, structures and the underwater environment must direct trained divers in a standard controlled way. All the information gathered during underwater inspection is not only of immediate value, but represents a milestone for all such work in the future life of the installation.

2.2 OPERATION AND MAINTENANCE OF IMPRESSED CURRENT CATHODIC PROTECTION SYSTEMS

J. L. TISCHUK, Occidental Petroleum (Caledonia) Ltd, Aberdeen, Scotland

ABSTRACT

The choice between impressed current and sacrificial cathodic protection systems is made at the platform design stage. Both types of system are capable of protecting the jacket from corrosion. Good performance from an impressed current system requires a high level of effort from both operating personnel and engineers. The operating histories of the impressed current cathodic protection systems on two North Sea platforms are described.

INTRODUCTION

Occidental's Piper and Claymore platforms are located in the North Sea, approximately 120 miles north-east of Aberdeen. The Piper jacket was installed in the summer of 1975 and the Claymore jacket was installed in the summer of 1976. Both were originally equipped with impressed current cathodic protection systems that featured platinum wire wound niobium anodes which were retractable through J-tubes. The anodes were supplied with current by transformer-rectifiers which used a pulsed current method to control current output.

In seven years of operation, the Piper CP system has required so many modifications and repairs that the only original components which remain are the copper transformer coils and the sections of the J-tubes below a depth of 120 ft. During this period, however, corrosion of the jacket has been minimal. The average depth of pitting measured on the submerged members is 1.5 to 2 mm.

The Claymore CP system received a major redesign on the basis of one year's operational experience with the Piper system. In six years of operation it has required many of the same modifications.

PIPER PLATFORM CATHODIC PROTECTION SYSTEM

The original design specification for the Piper cathodic protection system called for a current density of 8 milliamps per square foot (8 mA/ft^2). An impressed current system was chosen in preference to sacrificial anodes in order to save weight. In 1975, Piper was the heaviest jacket ever to be launched from a barge, and every means of minimising the jacket weight was used.

The original impressed current system consisted of 42 anodes and six Ag/AgCl reference cells. The anodes were 2 m long by 19 mm diameter niobium rods with a 0.2 mm diameter platinum-iridium alloy wire spot welded into a helical groove in the rod at a 3 mm pitch. The anodes were installed in J-tubes which were fabricated from 6 inch (152.4 mm), schedule 80 pipe and which distributed the anodes relatively evenly through the jacket. The J-tubes terminated in 3 m long fibreglass receptacles which also acted as dielectric shields. The anodes were supplied with current by four transformer-rectifiers, each of which had 1300 A output capacity at 36 V d.c. The feeder cables to the anodes were run in open cable trays to the +68 ft level and in standard electrical conduit down the legs and along the +20 ft level to junction boxes at the top of the J-tubes.

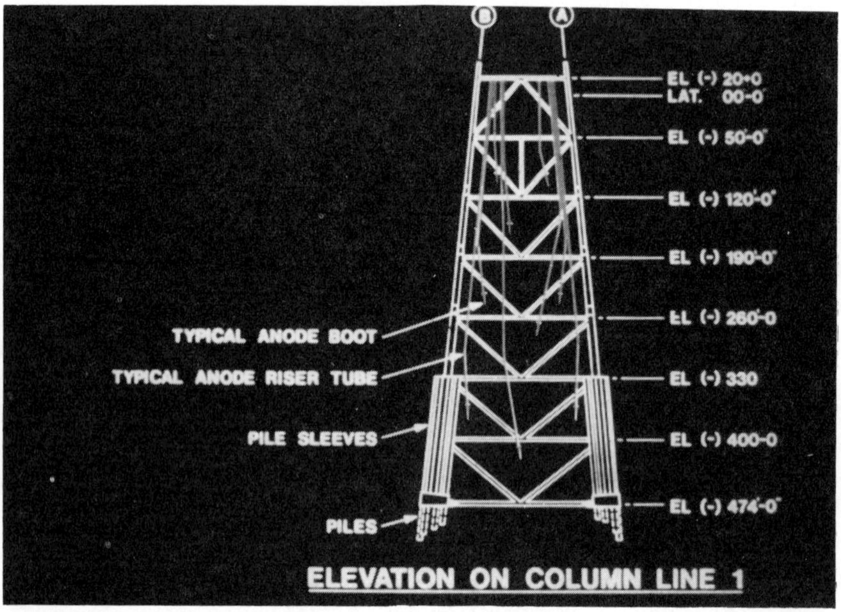

Fig. 2.2,1 – Schematic of Piper impressed current system.

During the winter of 1975–1976 a significant portion of the electrical conduit at the +20 ft level was washed away by high seas. Very little was done during 1976 due to other construction priorities. During the winter of 1976–1977 the remaining conduit and several of the J-tubes were lost due to storm damage. A remedial programme was begun in 1977 which included running armoured cable from the +68 ft level to the tops of the J-tubes, replacing the missing J-tubes and fitting additional clamps to secure the remaining J-tubes to the jacket. A full potential survey was not done on the jacket, but scattered potential measurements by divers indicated that the jacket had not yet polarised.

Fig. 2.2,2 – Typical wave action on Piper platform.

By the spring of 1978, 16 anode J-tubes were missing, 10 were still in place but had no anodes installed and 14 J-tubes had new anodes with armoured cables installed. Three of the six reference cell risers had been lost and the other three reference cells were out of service due to cable damage. In addition, two of the transformer rectifiers were out of operation due to electrical failures, some of the electrical wiring had been damaged due to other construction work, and a few of the anodes had failed due to excessive platinum consumption rates.

Because of all of these problems, a major cathodic protection retrofit was undertaken in 1978. The design review indicated that a current density of 8 mA/ft^2 was inadequate to polarise a North Sea structure and that 12 mA/ft^2 would be required. The final retrofit design called for fitting 530 sacrificial anode bracelets to protect the jacket between the splash zone and −150 ft. The anode bracelets weighed between 400 and 900 kg, nett, and were fitted to all horizontal, vertical diagonal and vertical members from −10 ft to −120 ft. Electrical contact was provided by four contact bolts per anode. The middle and bottom regions of the jacket were to be partially protected by 23 of the original impressed current anodes. This required fitting new sections of J-tube from −118 ft to +20 ft. The new sections were fabricated from 10¾-inch O.D. × 0.750-inch wall pipe and were held by clamps to the jacket members and were connected to the existing 6-inch J-tubes at −118 ft. Platinum wire wound niobium anodes with armoured cables were to be installed, with a cable joint on the legs at +26 ft and a further run of armoured cable up the legs to the +68 ft level.

The bottom region of the platform was also to receive supplemental protection from four buoyant anode arrays, each rated at 1000 A output. These anode arrays (trade name Solanodes) were placed on the sea-bed inside the jacket framework and floated 10 to 15 m above the sea-bed. Armoured feeder cables were run similarly to the other anode cables, down the legs, along the +20 ft level and down two risers to the sea-bed. Each anode array was supplied with current by individual Thyristor controlled transformer-rectifiers rated at 1000 A output at 60 V.

By 1 January 1979, 390 of the sacrifical anodes had been installed, all 23 J-tubes had been repaired, 18 impressed current anodes had been installed and were operating, and two of the anode arrays were installed and operating. The remaining two anode arrays were installed in January 1979, and the remaining sacrificial anode bracelet anodes were installed during the course of 1979.

By June 1979, 11 of the 18 impressed current anodes had failed due to loss of the platinum wire from the niobium rods. The anodes were refurbished by electroplating with 15 microns thickness of platinum and reinstalled. Anode failures continued and laboratory work revealed that a short duration negative spike which was a characteristic of the pulsing transformer-rectifier output was responsible for a 30-fold increase in the platinum wear rate of the impressed current anodes. As an interim measure to prevent further anode failures, the pulsing control circuits on the transformer-rectifiers were bypassed in order to eliminate the negative spike.

A thorough cathodic protection potential survey was conducted on the jacket in November–December, 1979. During the survey 22 of the 23 impressed current anodes were operating at 150 to 200 A output each and the four Solanodes were operating at 700 A output.

The results of the survey indicated that the jacket had been polarised and except for a few isolated readings was completely protected. The survey also showed that the areas near the Solanodes were over-protected and the Solanode outputs were adjusted accordingly.

On 1 January 1980 the Piper CP system was formally handed over from the Construction group to the Offshore operating group. The completion of the major retrofit did not bring the end of all problems. The system had no reference cells and only the Solanodes were controllable. The control circuits on the original transformer-rectifiers were still bypassed. A project was undertaken to modify the transformer-rectifiers. It was May 1981 before the modifications were carried out, i.e., by the time that funding was established and resources allocated. The major cause of downtime in the system since 1980 has been cable faults between the +68 ft level and the J-tubes. In order of decreasing frequency the cable faults have been due to:

1. Water ingress into heat shrink insulation at joints.
2. Wear due to cables not being firmly clamped into position.
3. Accidental damage from other construction activities.

During late 1981 all cable runs were made secure to the platform members and all heat shrink insulation was replaced with epoxy coatings. This has significantly reduced the occurrence of cable faults.

Additional potential surveys in 1980, 1981 and 1982 have indicated that the Piper platform is well protected. By reducing the current outputs from the sea-bed anode arrays, we have substantially reduced the over-protection in the lower regions of the jacket.

Our most serious problem at present is damage to the Solanodes. About ten per cent of the anode rods and many of the non-conductive anode supports have been broken off by falling debris, which is mainly dropped scaffolding. In addition, one of the feeder cables has a subsea fault. The damage has not affected the ability of the system to protect the jacket, but we are planning a programme of remedial works to repair the damage.

Underwater close visual inspection of the Piper jacket has revealed some pitting corrosion on all members examined. The average pit depths are 1.5 to 2.0 mm and the deepest pits are no more than 3.0 mm deep. These corrosion pits do not appear to have become deeper or more extensive since 1979.

2.2,3 – Typical corrosion pitting in the jacket.

CLAYMORE PLATFORM CATHODIC PROTECTION SYSTEM

Because of the early problems with the Piper CP system, the original design of the Claymore CP system was modified while the jacket was still in the fabrication yard. The revised design was based on a current density of 15 mA/ft^2 for steel exposed to sea-water and 2 mA/ft^2 for steel buried in mud. The J-tubes were upgraded from 6-inch schedule 80 pipe used on Piper to 10-inch schedule 120 pipe for the first 120 ft on Claymore, reducing to 6-inch schedule 80 below the −100 ft level. Three-inch pipe was used for electrical conduit at the +20 ft level and 6-inch pipe was used to protect the cable runs down the legs from the +68 ft level to the +20 ft level.

The same platinum wire wound niobium anodes and pulsing current controlled transformer-rectifiers were used on Claymore as had been used on Piper. A total of 55 anodes and 12 reference cells were distributed through the jacket.

2.2,4 – General arrangement of the Claymore impressed current system.

The Claymore jacket was launched in June 1976. One of the J-tubes was damaged during the launch. In October 1977 the cathodic protection system was energised. The commissioning report indicated that all 55 anodes checked out and were working properly. The damaged J-tube had not been repaired.

The first potential survey of the Claymore jacket was conducted in September 1978. Thirty-six of the 55 anodes were in working order. About 90 per cent of the jacket was polarised to protected potentials. The underprotected areas were typically near inoperational anodes.

A programme was undertaken in 1979 to replace the inoperational anodes. During this same year, all anode feeder cables were replaced from the transformer-rectifiers to the tops of the J-tubes. The rewiring was necessary because of damage and deterioration of the cable insulation and corrosion of bolted connections inside the junction boxes.

A potential survey was done in September 1979. During the survey 45 of the 55 anodes were operating. The protection of the jacket was very much improved over the first survey. The underprotected areas were fewer in number and smaller in area, although there were some areas of overprotection.

Up to mid-1979 when the rewiring of the Claymore CP system took place, the main cause of anode failure was damage due to falling debris. After the rewiring, the anodes began to experience the premature platinum loss which occurred previously on Piper.

The transformer-rectifiers were modified to eliminate the pulsing in November 1980, but much of the damage had been done by then. A potential survey was carried out in December 1980, during which 27 of the 55 anodes were operating at their maximum available output. There were virtually no areas of underprotection, although a certain amount of overprotection was accepted near the operating anodes.

New anodes were ordered for Claymore in January 1981. These anodes were metallurgically bonded platinised niobium, and were delivered in early September. By the time we had attached cables to the anodes and begun shipment offshore, it was October and only 19 of the original anodes were operating.

By December 1981 we had 30 new anodes operating and after getting diving assistance to remove various objects from the bottoms of the J-tubes we had 49 anodes operating in July 1982. It was not possible to install the remaining six anodes because three J-tubes had been removed from the jacket because of fatigue problems with their attachments, two J-tubes had been removed and replaced as part of another job and the replacement work had not been properly done, and the J-tube that was missing after the launch of the jacket had not yet been replaced.

Underwater inspection has shown similar patterns of pitting corrosion to Piper. The extent and depth of pitting is slightly less at 1 mm average and 2 mm maximum pit depth.

SUMMARY

Despite a number of original design faults, the impressed current cathodic protection systems on Piper and Claymore have protected both jackets from serious corrosion. Both systems required several million dollars worth of refurbishment to improve their reliability and both still require thousands of man-hours each year to keep them operational.

2.3 MIXED METAL GALVANIC PROBLEMS ASSOCIATED WITH OFFSHORE SEA-WATER PUMPING EQUIPMENT

D. N. C. ROBSON, Britoil Ltd, Aberdeen, Scotland

ABSTRACT

An offshore sea-water pumping system consists of distinct processing stages in a logical sequence; each of these is considered from the point of view of material selection for vessels and equipment, with particular emphasis on the avoidance of galvanic corrosion problems which commonly arise in such systems. The reasons for many of the problems presently encountered are highlighted and cost effective means of overcoming them described. The five main process stages looked at in detail are:

(i) Raw sea-water supply pumps.
(ii) Heat exchange with other process fluids.
(iii) De-aeration.
(iv) Filtration.
(v) Sea-water injection.

Sea-water pipework is considered from the point of view of its interaction with the vessels and equipment comprising these five basic stages, but is not considered in detail.

GENERAL OUTLINE OF AN OFFSHORE SEA-WATER PUMPING SYSTEM

An offshore sea-water pumping system is composed of five principal stages: (i) Sea-water lift pumps; (ii) heat exchange; (iii) de-aeration; (iv) filtration and (v) injection. Below is a brief outline of the type of equipment used in each of these stages.

(i) Sea-water lift pump

On steel jacket structures there are two options for this equipment:

(i) Motor drive at production deck level driving a long shaft to which are fitted the impellers of a multistage pump located near the bottom of the caisson.
(ii) Motor drive and multistage pumps in one complete cartridge which is suspended from the rising main with power supplied via a cable clipped to the exterior of the rising main.

(ii) Heat exchange equipment

Generally, incoming sea-water exchanges heat with the cooling medium (usually fresh-water) of a platform and the produced oil flow, prior to de-aeration. On some platforms a proportion of the pumped sea-water is used as the cooling medium. Heat exchange equipment is often in the form of plate type heat exchangers, as this type of equipment takes up a lot less space than the tube-in-shell exchanger. However, in many instances, for operational reasons, tube-in-shell exchangers are installed.

(iii) De-aeration equipment

There are two options open for this stage of an offshore sea-water circulation system:

(i) Vacuum de-aeration.
(ii) Gas stripping (generally using hydrocarbon gases separated from the produced oil).

(iv) Filtration equipment

Sea-water filtration equipment on an offshore production platform generally consists of either packed beds or polypropylene mesh filters, although on some platforms the process sequence will involve filtration of the sea-water prior to its de-aeration.

(v) Sea-water injection pumping equipment

Generally, high speed multistage centrifugal pumps are used. The usual materials of construction are 316L stainless steel or duplex austenitic steel alloys for both pump casings and impellers.

Other sea-water pumping systems on a platform include:
(a) Fire-water pumping systems. These are similar to the continuously operating sea-water lift pumps, but operate only when the fire-water system is used. On some platforms these systems also supply cooling water to the auxiliary diesel generators.
(b) Hypochlorite generating unit. This normally consists of banks of electrolytic cells operating in parallel. The pumps, valves and pipework associated with this equipment are generally more prone to corrosion problems than other similar equipment handling raw sea-water mainly because even small traces of free chlorine in sea-water will significantly increase its corrosivity.

GALVANIC PROBLEMS ASSOCIATED WITH VARIOUS PARTS OF A SEA-WATER PUMPING SYSTEM

(i) Sea-water lift pumps and fire-water lift pumps

Usual materials of construction for this equipment are as follows:

(i) Pump casings – Ni-resist (e.g., AUS 202 Grade B), duplex alloys (e.g., Zeron 25, Feralium Rex 734, etc.).

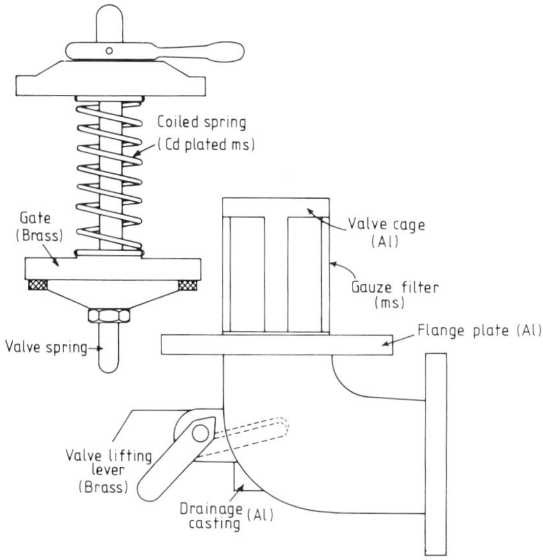

Fig. 2.3, 1 – Valve used in sea-water systems (note number of materials required).

Fig. 2.3, 2 – Actual valve (note corrosion of mild steel, cf. no corrosion of valve body).

(ii) Impellers – Ni-resist, stainless steel 316L, duplex austenitic alloys.
(iii) Shafting – Stainless steel, Monel, duplex alloys.
(iv) Rising main – Lined carbon steel, Ni-resist, Al-bronze, 316L stainless steel.

Corrosion of sea-water/fire-water lift pump casings and impellers. Galvanic problems can arise when stainless steel impellers are installed in Ni-resist casings because although this is generally acceptable design philosophy (i.e. pump casing constructed of a material which is anodic to that of the impeller) the difference between their respective electrochemical potentials in sea-water causes rapid corrosion where the impeller meets the shaft. An aggravating factor with sea-water lift pumps is the geometry of the design where the ratio of the internal surface area of the casing to that of the impeller is not sufficiently large to reduce the galvanic corrosion rate on the surface of the Ni-resist to acceptable levels. Ideally, the impeller and shaft should be constructed of the same material.

Yet more galvanic problems are associated with the wear rings. Thus, the casing wear ring needs to have a lower Brinell hardness than the impeller or impeller wear ring and for this reason bronzes were often used. However, many bronzes are anodic to the cast irons and Ni-resist cast irons used for pump casings and this, of course, is most unsuitable since the surface area of the wear ring is small compared to that of the casing. As a result, such an item would be subject to quite rapid corrosion.

It is important, therefore, that wear rings be made of materials that are galvanically more noble than are the impellers and casings with which they are associated. The nickel/copper alloy 506 and stainless steels perform better than the bronzes in this service.

When the pump casing is constructed of stainless steel or duplex alloy and the impeller is similarly constructed, it is often quite useful to fit impeller wear rings of specially hardened stainless steels or non-galling duplex alloys (e.g., 17/4 PH stainless steel, Nitronic 50, Nitronic 60), since a hard, galvanically noble wear ring on the impeller ensures that it does not wear down and unbalance the impeller and shaft assembly.

Corrosion of rising main. Galvanic corrosion of the rising main can be avoided by constructing it of the same material throughout and by giving due consideration to the material of construction of the submersible pump casing and discharge head to which the rising main is connected. If possible, these should be constructed of the same material as the rising main.

It should be remembered that for a 12 in (305 mm) diameter rising main the internal surface area of the pipe, to a distance of 2–3 ft (610–915 mm) from the end, will act on the internal surface area of the discharge head or final delivery chamber of the submersible pump. Therefore, the internal surface area of rising main liable to have a galvanic effect on either the discharge head or pump casing is likely to be greater than the internal surfaces of either of these two items. Thus, clearly, it is imperative that the discharge head or pump casing should be constructed either from the same material as that of the rising main or one that is more noble.

For instance, if the rising main is constructed of stainless steel then neither the discharge head nor the submersible pump casing should be constructed of cast iron Ni-resist or aluminium bronze but should instead be stainless steel or an austenitic duplex alloy.

If for some reason the material of construction of the discharge head or submersible pump casing cannot be made the same as, or more noble than, that of the rising main (e.g. due to non-availability of these items in the requisite material), then the sections of rising main to which these parts are attached should be internally coated with solventless epoxy or solventless urethane coatings to thicknesses of at least 500 microns. Coating of stainless steels or duplex alloys requires extreme care as most coatings do not adhere well to these materials. The above-average hardness of stainless and duplex alloys means that either aluminium oxide grit or garnet has to be used for surface preparation to achieve a suitable profile for the application of solventless epoxies. Use of such abrasives also avoids the problem of rust specks on stainless steel when chilled steel grit is used and which is probably caused by galvanic action between the stainless steel and the particles of steel grit which remain on the surface after grit blasting operations.

Galvanic corrosion of internally coated carbon steel rising mains. In the past, to reduce the costs of sea-water pumping equipment, internally coated carbon steel rising mains have been installed on some platforms. These are prone to very serious galvanic corrosion as small defects in the coating in the vicinity of the discharge head or pump casing will cause rapid penetration of the pipe wall by the sea-water. This occurs because these items of equipment are invariably constructed of a material more noble than carbon steel, and thus a galvanic cell can arise in which the anodic part (i.e. the carbon steel rising main) has a far smaller area (i.e. the coating defect) than the cathodic part (i.e. the internal surface areas of the discharge head or submersible pump delivery chamber).

On one platform, the design engineer thought that he could have the best of both worlds by using internally and externally coated carbon steel pipe, flanged at each end with stainless steel flanges. He clearly thought that he could have a low cost system and yet have no corrosion in the areas where effective coating was not possible. The end result though was that the inevitability of coating break-down in the vicinity of the flanges led to very rapid penetration of the pipe wall at these areas of coating break-down, due to the galvanic effect of the stainless steel flanges on the small areas of exposed anodic carbon steel.

It is clear from the above, therefore, that if an internally and externally coated carbon steel rising main is to be used, the quality control of the application of the coating has to be devoted to ensuring that edges and corners are

Fig. 2.3, 3 – Corrosion of carbon steel butterfly valve in brass alloy Yorcalbro valve body.

Fig. 2.3, 4 – Overall view of butterfly valve in diesel generator cooling system.

ground to have a radius of at least 1/8 in (3.2 mm) to ensure that the coating can be applied at a sufficient thickness. Furthermore, application of conventional coatings will have to be by stripe coating in these areas. I would recommend either a two-component solventless epoxy or a urethane for this service, as both can be applied to substantial thicknesses in one coat.

(ii) Galvanic corrosion of heat exchange equipment in sea-water pumping systems

There are a number of different types of heat exchange equipment used; titanium plate heat exchangers, stainless steel tubes and tube-plates in carbon steel shells with sea-water on the tube-side, cupro-nickel tubes (90:10) and tube-plates in carbon steel shells again with sea-water on the tube-side, are but a few of the different types of heat exchanger which can be used.

The heat exchange equipment in a sea-water circulation system performs a number of functions such as cooling a platform's closed loop cooling water system, cooling produced oil and gas, condensation of produced gas and intercooling between gas compression stages.

Galvanic problems associated with titanium plate heat exchangers. Since lined carbon steel pipe is often used in sea-water systems, the galvanic effect at titanium plate heat exchangers is particularly severe as, once again, small areas of coating break-down near the sea-water inlet or outlet flanges expose small areas of anodic carbon steel to the relatively large area of titanium plate which can have a galvanic effect on them, and corrosion of the pipe wall rapidly ensues. It should be noted at this juncture that electrical insulation of the exchanger from the pipework or vice versa should not be relied upon, as the electrical insulation of even the most meticulously installed gaskets and grommetted nuts and bolts eventually deteriorates due to external contamination by dirt and moisture. Furthermore, it would require continual checking and maintenance of the electrical insulation to ensure that the pipework and exchanger remained electrically isolated from one another. This constitutes additional maintenance work which a platform's maintenance team can well do without.

A further difficulty with electrically insulating equipment from the associated pipework is that even when there is effective insulation at a flange, low resistance current paths are available between the exchanger and its associated pipework via supports and brackets, etc. These are electrically continuous with the deck and platform structure. Even were these made electrically insulating, it is likely that after the first dismantling operation, much of it would go by the board.

The solution is to install an intermediate 316L stainless steel spool between the exchanger and the pipework, the galvanic potential of 316L stainless steel being nearly the same as that of titanium. The stainless steel spool would nonethe-less have to be internally coated and electrically insulated at the flanged joints where it joins to the exchanger at one end and the lined carbon steel pipework at the other. This is because the backing plate on a titanium plate heat exchanger is often lined carbon steel and it is necessary to reduce the surface area of the cathodic stainless steel which can have a galvanic effect on small areas of coating break-down on the lined carbon steel backing plate, and also the lined carbon steel pipework. The comments made in a previous paragraph regarding coating of stainless steel clearly apply to this application as well. Electrical insulation at flanges in this instance is simply a back-up in that it increases the resistance of the current path between the stainless steel and the lined carbon steel. This then is the philosophy which should be adopted towards electrically isolating flanges. They should never be completely relied upon to stop corrosion, but can serve a very useful purpose as an additional safeguard against galvanic corrosion.

Galvanic problems associated with tube-in-shell exchangers. The most common problem is that of a lined carbon steel waterbox corroding at areas of coating break-down due to the galvanic effect of a tube-plate constructed of a more noble material.

This problem occurs in heat exchangers where the sea-water is on the tube-side and the fluid being cooled on the shell-side. The two methods of overcoming this are:

(1) Installation of sacrificial anodes in the waterbox. (It should be noted, however, that a flake of paint extending over an area where the water has penetrated can often prevent the cathodic protection reaching the steel.)
(2) Constructing the waterbox of the same material as that of the tube-plate or constructing it of a more noble material and then coating it to reduce its galvanic effect on the tube-plate.

(iii) Corrosion of de-aeration equipment

De-aeration of sea-water on a platform is carried out either by vacuum de-aeration or by gas stripping. It generally occurs after heat exchange with produced oil and the closed loop cooling water system – where it exists – of the platform. De-aeration of the sea-water can occur either before or after the filtration stage. One of the advantages of de-aerating the sea-water prior to filtration is that the sea-water becomes less corrosive after de-aeration, and hence corrosion of filtration equipment will be less severe, although galvanic action still has to be guarded against even at oxygen concentrations of less than 10 p.p.m. Galvanic corrosion problems associated with de-aeration equipment are generally centred on control valves, spray heads and pumps.

(iv) Corrosion of filtration equipment

Filtration equipment often consists of polyester flake glass-lined carbon steel

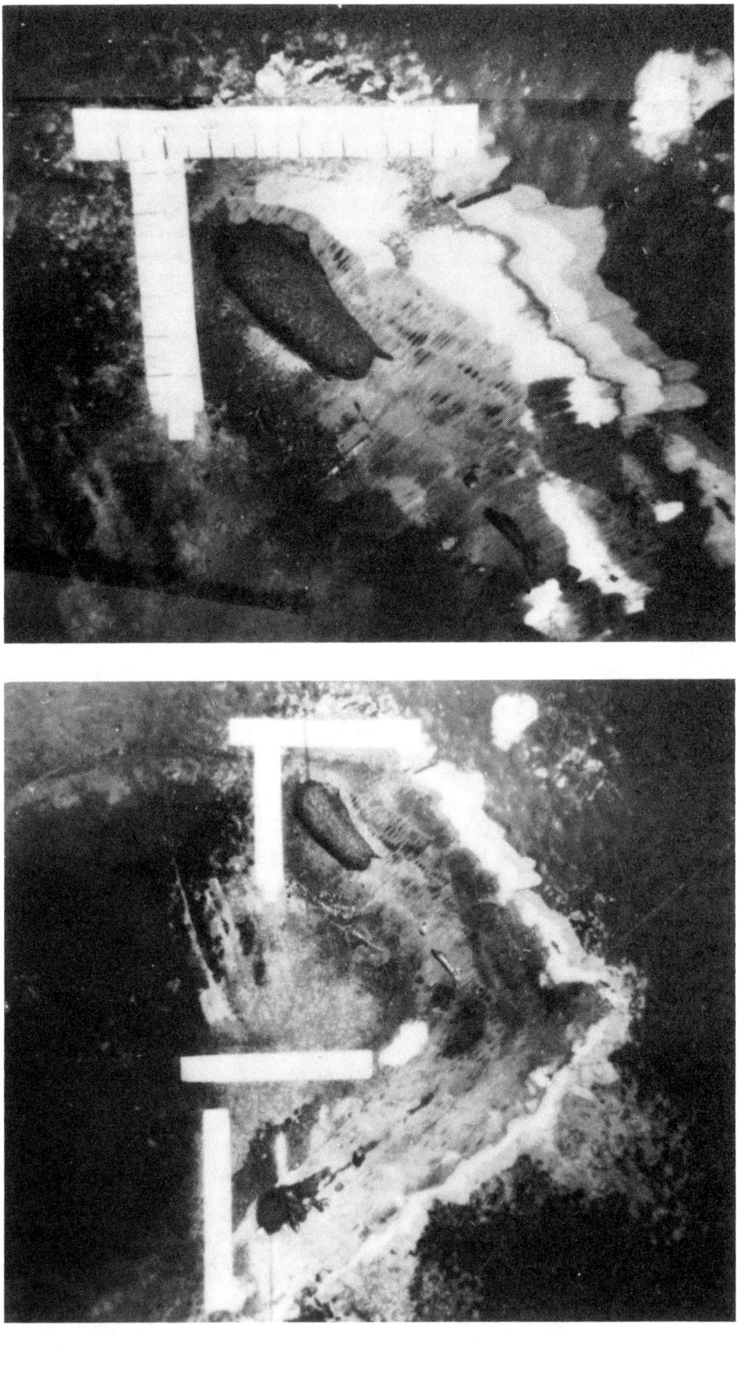

b

a

Fig. 2.3,5 – (a) Galvanically initiated corrosion where a stainless steel distributor and manifold was resting on a support in a sandfilter. (b) Enlargement of part of (a).

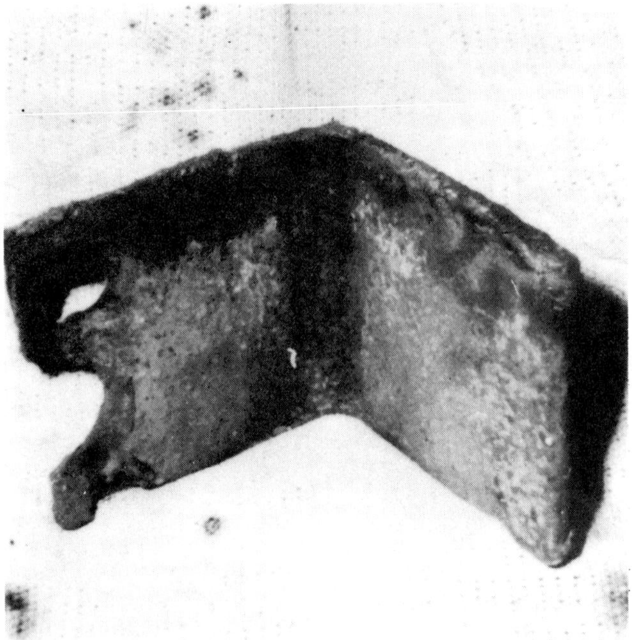

Fig. 2.3,6 – Carbon steel support on which stainless steel manifold rested.

vessels filled with anthracite, garnets and sand, with the sea-water flow entering the vessel via multi-branched sparger pipes.

One area which is particularly prone to galvanic problems is the inlet nozzle, particularly if the water distributors are constructed of a corrosion resistant material such as stainless steel. This is because the relatively large internal surface area of the distributor can have a strong galvanic effect on small areas of coating breakdown within the inlet nozzle. The internal surfaces of the nozzle moreover, are particularly prone to coating breakdown for two reasons: first, it is difficult to apply an adherent, defect free, coating to the inside of a relatively small diameter pipe, even to one with a diameter of as much as 6 in. Secondly, there is often considerable turbulence and cavitation in an inlet nozzle caused by control valves, changes in pipe section or rapid changes in direction of fluid flow.

This problem can generally be overcome by internally lining the inlet nozzle with the same material as that of the distributor. While not ideal, it does avoid the problem of having to try to weld, say, a stainless steel nozzle into a carbon steel vessel.

It is probably a good idea to coat internally such a nozzle to reduce the surface area of cathodic material which can act on an area of exposed carbon steel where the internal coating has broken due to cavitation or erosion.

(v) Galvanic problems associated with sea-water injection pumping equipment
Due to the compressed geometry of multistage centrifugal sea-water injection pumps and the high turbulence of the fluid flow through them, the casings are generally made of stainless steel or duplex alloys. The impellers are similarly constructed, even though the sea-water is oxygen-free at this stage in the process.

The problems associated with sea-water injection pumping equipment are the classical ones of galvanic action between shaft and impeller, impeller and impeller wear ring, casing and casing wear ring, etc.; where the material of construction has been chosen without due regard to galvanic factors.

In taking galvanic effects into account in choosing materials for pump components, special attention must be given to the varying galvanic relationship between stainless steels and the nickel and copper alloys that may be associated with stainless steels in pumps. In contact with moving sea-water the stainless steels are in their passive condition, exhibit a nobler potential than do the nickel and copper alloys and are given some galvanic protection by them. On the other hand, stainless steels that develop areas of local activity or loss of passivity in quiet or stagnant sea-water or in crevices within bearings or under packing are less noble than the copper alloys and are not given galvanic protection by them.

In pumps, where such galvanic protection of stainless steel components is vital, the salt-water in the pump should be replaced by flushing and filling with fresh-water during periods of prolonged shutdown. This practice should be followed also with pumps in which all the principal components are stainless steel, since the surfaces of stainless steel that remain passive can cause galvanically accelerated corrosion of the same, or similar, alloys that become active in pits, under deposits, or within crevices.

It should be remembered that cast iron corrodes and develops a layer of graphitic corrosion products which is more noble than the cast iron from which it originates. Consequently, a cast iron casing should not be depended upon to provide galvanic protection to a copper alloy impeller or shaft and could even accelerate corrosion of such components. Economic factors lead to the consideration of such materials as the austenitic cast irons (generally referred to as Ni-resist) for pump casings. The Ni-resist austenitic cast irons are as much as ten times more resistant to high velocity effects than unalloyed or low alloy content cast irons of either the flake graphite or spheroidal graphite (ductile cast iron) types. The austenitic cast irons are galvanically more noble than plain cast iron, but less noble than bronzes or passive stainless steels. Their galvanic potentials are closer than those of bronzes to the potentials of stainless steels in the active condition. Consequently, austenitic cast iron casings have an advantage over bronze casings when used in conjuction with stainless steel impellers and shafts during shutdowns where crevice corrosion and pitting of the stainless steels may be promoted by the galvanic effect of the bronzes.

In conclusion, it can be seen from the above that an offshore sea-water circulation system is composed of distinct stages and the criteria for materials selection differ for each stage, with the most stringent requirements occurring with equipment handling aerated sea-water. The high cost of maintenance offshore favours the selection of corrosion resistant materials over lined carbon steel, which carries with it a high risk of coating breakdown. However, even with corrosion resistant materials, care has to be exercised in the design of equipment to avoid crevices and unsuitable anodic/cathodic area ratios.

2.4 CATHODIC PROTECTION AND STEEL PROPERTIES

T. G. GOOCH, The Welding Institute, Abington, Cambridge

ABSTRACT

The general effect of hydrogen in causing embrittlement of ferritic steels is well recognised. The paper considers the extent to which hydrogen pick-up from cathodic protection of steel structures affects the material properties relevant to service. Attention is paid to steel grades employed in offshore oil and gas extraction, with reference to behaviour under both static and cyclic loading. The aim is to indicate the extent of current knowledge, and define potential problem areas where further data are necessary.

1. INTRODUCTION

The use of cathodic polarisation to prevent corrosion of steel in offshore service means inevitably that hydrogen ion discharge takes place on the metal surface to a greater or lesser extent. In consequence, the possibility of hydrogen pick-up by the steel must be recognised and attention given to the effects of hydrogen on the mechanical properties of the material. The degree of practical problem posed depends on the amount of hydrogen entering the steel, on the sensitivity of the steel to hydrogen embrittlement, and on the operative stress system. By and large, the variables likely to have the major influence on the overall mechanical integrity of a structure have been identified, but full quantification of these factors has not been achieved: the paper aims to summarise available information and indicate where further work is required.

2. HYDROGEN PICK-UP

In essence, hydrogen discharge from the environment takes place on the steel surface in two stages [2.4,1] namely

$$H^+ + e^- \rightarrow H \qquad (2.4,1)$$

$$H + H \rightarrow H_2 \qquad (2.4,2)$$

Both reactions are promoted by more negative cathodic potentials. Cathodic protection systems are conventionally designed to maintain a potential sufficiently negative to suppress corrosion of the steel, without necessarily leading to major gas evolution, i.e. polarisation to below about −0.82 V with respect to a saturated calomel reference electrode (SCE). Apart from minimising current demand, the aim is to minimise material hydrogen pick-up associated with a high cathodic current density, and avoid deleterious effects of over-protection on any protective paint system.

It would thus be expected that the first variable determining the amount of hydrogen pick-up would be the cathodic protection potential. In principle, the degree of embrittlement depends on the local hydrogen level attained at cracks or similar defects in areas of tensile stress, but this localised concentration is virtually impossible to determine and reference must be made to bulk hydrogen contents. Bearing in mind the significance attached to material hydrogen level in other situations where hydrogen-induced embrittlement and cracking can occur, (for example, arc welding [2.4,2], surprisingly few data exist to illustrate the effect of potential on hydrogen pick-up by structural or similar steels in sea-water conditions. At least with unstressed material, and possibly with static loading, it is unlikely that polarisation as normally applied will cause marked ingress of hydrogen (Table 2.4,1) [2.4,3]. However, additional factors should be recognised. First, both anodic and cathodic corrosion rates are increased by cyclic loading [2.4,4] and it seems clear that the rate of hydrogen absorption is enhanced on metal undergoing strain, especially if localised plastic deformation occurs exposing clean material to the environment; this can be particularly significant with more negative potentials as considered below. In fact, elastic strain can also promote hydrogen pick-up, particularly in materials of harder microstructure [2.4,4] [2.4,5]. Secondly, hydrogen pick-up will be enhanced if reaction 2 is poisoned [2.4,1] leading to 'build-up' of adsorbed hydrogen atoms on the metal surface from reaction 1.

Table 2.4,1

Results of hydrogen analyses on unstressed samples exposed to different environmental conditions [2.4,3]

Environment	Sample type	Exposure time (h)	Total hydrogen level (ml/100 g)
(Air)	BS 4360 : 50 D	0	0.06
	Weld metal	0	0.05
Sea-water with cathodic protection by Zn at +20 °C	Weld metal	1000	0.05
		4000	0.05–0.08
Sea-water with 100 p.p.m. H_2S	Weld metal	350	0.08
		2500	0.02–0.2
NACE	BS 4360 : 50 D	100	8.3
		525	7.8
	Weld metal	100	1.1–2.7
		525	0.5–3.0

This latter situation can arise in the presence of various agents, most notably hydrogen sulphide. Concern has arisen over the possibility that sulphate-reducing bacteria (SRB) may be active in causing the local formation of H_2S on offshore constructions either under fouling or below the mud line. Indeed, closed container studies have shown that putrefaction of organic material in sea-water can lead to H_2S levels as high as 170 p.p.m. [2.4, 6]. It is, however, not clear that the conjoint action of cathodic protection and SRB will necessarily lead to unacceptably high hydrogen levels in steel. Cathodic protection will engender an alkaline condition adjacent to the material surface, and in general this is not conducive to high SRB activity. Hydrogen-induced stress corrosion cracking (SCC) associated with H_2S from biological activity has been reported on offshore items [2.4,7], but to the author's knowledge, only at considerable depth (*ca* 400 m) below the mud-line under conditions where effective cathodic protection is most unlikely to have been applied.

3. MATERIAL VARIABLES

In terms of the effects of material composition and microstructure on susceptibility to hydrogen embrittlement, the major factor is the material hardness and strength level. Hardened structures are deleterious, and the risk of environmental hydrogen cracking is very much greater in quenched material than in annealed or normalised steel [2.4,8].

Because of the overriding role of hardness, concern attaches to welded joints in view of the transformation and hardening promoted by the welding thermal cycle. Considerable control can be exerted over the hardening experienced in weld areas by appropriate selection of steel composition and adjustment of welding conditions to promote slow cooling from peak temperature [2.4, 2]. Since from the practical standpoint, risk of hydrogen embrittlement in relatively soft parent material is negligible, it becomes an attractive concept to stipulate a maximum hardness which will not be exceeded at welds so that hydrogen embrittlement from environmental effects is avoided [2.4, 9]. Practically, however, this is difficult because some hardness increment will virtually always occur with normal fusion welding procedures, and this is true even if post weld heat treatment is applied. There is no abrupt change in material susceptibility to hydrogen embrittlement with increasing hardness, and the problem thus becomes one of defining what hardness level is tolerable for a given offshore service application. Various guidelines have been propounded, ranging typically from 250 HV to 325 HV, but these represent 'engineering judgements' rather than results of definitive test programmes: they are in effect a compromise between the exigencies of fabrication and the NACE Rockwell 22 C (243 HV) criterion widely accepted as the maximum permissible hardness to avoid stress corrosion in severe H_2S conditions [2.4,10]. Published information indicates that bulk hydrogen uptake in sea-water with cathodic protection at least to about

$-0.9\ V_{SCE}$ is substantially less than in saturated H_2S media (Table 2.4, 2). Hence, specified hardness limits based closely on the Rockwell 22 C level will be unnecessarily conservative for normal marine service.

Table 2.4, 2
Results of SCC tests on undressed welds in chloride media [2.4, 9]

Material	Environment	Parent yield stress σy (N/mm^2)	Maximum HAZ hardness HV 2.5	Stress level (% σy)	Test duration (h)	Comments
NCMV	3.5% NaCl	1630	485	67	550	Failure
			506*	67	1.3	Failure
En 40 C	3.5% NaCl	1150	463	67	2370	No failure
		1150	620*	85	1483	Failure
BS 4360 : 55 E	Simulated sea-water	>415	455	50	1000	No failure
			455	85	1000	No failure
			455	100	1000	No failure
			455	100	3530	No failure
BS 4360 : 50 D	Simulated sea-water cathodic protection at −0.9 V relative to Ag/AgCl electrode	>345	473	100	3280	No failure
			473	100	9860	No failure

*Weld metal.

Moreover, blanket imposition of a single hardness level must be regarded with caution. Most published data on the microstructure dependency of hydrogen embrittlement pertain to medium carbon or alloy steels, having been derived from investigations on very high strength grades (yield stress >1000 N/mm^2) or into the problem of H_2S induced SCC [2.4, 8]. From such studies, microstructure *per se* clearly is important, sensitivity to hydrogen increasing markedly if a twinned martensite structure is developed, for example. Rather less information exists for microstructures appropriate to structural or pipeline steels, but it would seem that finer transformed microstructures are to be preferred at hardnesses of say 200-280 HV [2.4,11] [2.4,12], and recent work on weld metal hydrogen cracking during fabrication has shown [2.4,12] increased cracking resistance in microstructures of higher toughness specifically at low bulk hydrogen contents (<5 ml/100 g), as might be expected with pick-up from cathodic protection.

Resistance to hydrogen cracking depends also on material composition [2.4,13], including residual element contents, with clean steels being preferred [2.4,14]. From the practical standpoint, this is probably secondary to that of hardness, but further work on composition and microstructure would seem warranted, especially in view of recent developments in steel-making practice, aimed at producing materials of low total alloy content to reduce hardening and the risk of hydrogen cracking during welding.

4. STATIC LOADING

Although not truly static loading, it is well recognised that tensile ductility can be greatly reduced by cathodically introduced hydrogen. Most tensile tests have been carried out in air on previously charged samples, and have employed either high strength steels or severe charging conditions to demonstrate the marked mechanical effects that can stem from hydrogen [2.4,15]. Little attention has been paid to structural or similar steels following cathodic polarisation in sea-water, but it is unlikely that simple loss of tensile ductility would be practically significant. A more pronounced effect on ductility is evident [2.4,16] [2.4,17] under slow strain rate testing (Fig. 2.4,1); with simultaneous cathodic polarisation in sea-water, even parent structural and pipeline grades can show appreciable loss of ductility and cleavage fracture, presumably because hydrogen ion discharge and hydrogen pick-up are enhanced by the dynamic strain experienced leading to high hydrogen contents, at least locally within the material.

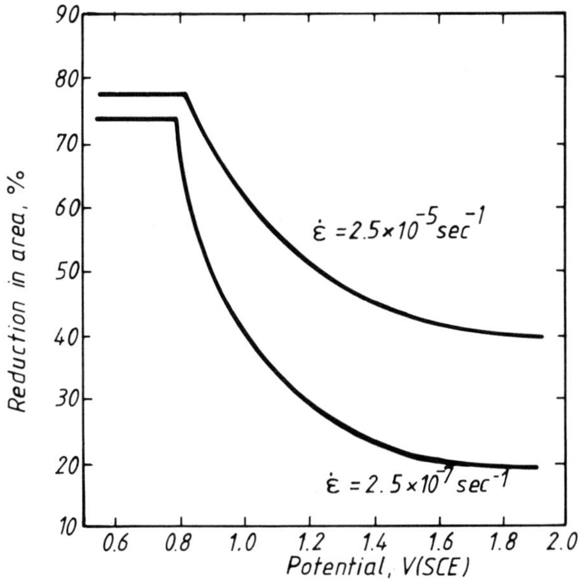

Fig. 2.4,1 – Variation in tensile ductility with cathodic polarisation of X65 line-pipe steel from slow strain rate tests in 3.5% NaCl of pH 8 (from Ref. [2.4,16]).

At present, the practical import of this effect is not clear. It shows that even in soft material, cathodic polarisation can promote discernible hydrogen embrittlement, over-protection reducing ductility to only 15 per cent RA, but there seem to be no instances of service failure which can be directly ascribed to such a combination of stress and environmental variables. This may be because the particular combination of test parameters necessary to induce low ductility represent a statistically unlikely situation in practice, possibly with intermittent hydrogen diffusion away from areas undergoing tensile strain during 'hold periods' of the service loading cycle. Bearing in mind the extensive use of stressed steel in sea-water over many years, it would seem improbable that the effect constitutes a major practical hazard, but some additional study is necessary, particularly to examine the response of harder microstructures associated with welds to conjoint tensile strain and cathodic protection.

A number of investigations have studied the effect of saline conditions in promoting cracking under constant load using precracked samples and fracture mechanics analysis of the results. From these programmes, an environmental effect even without cathodic polarisation can be detected [2.4, 9] to low hardness levels, e.g. *ca* 200 HV. Less attention has been paid to cathodic protection, but, as illustrated by Fig. 2.4, 2, this increases the risk of cracking, especially at

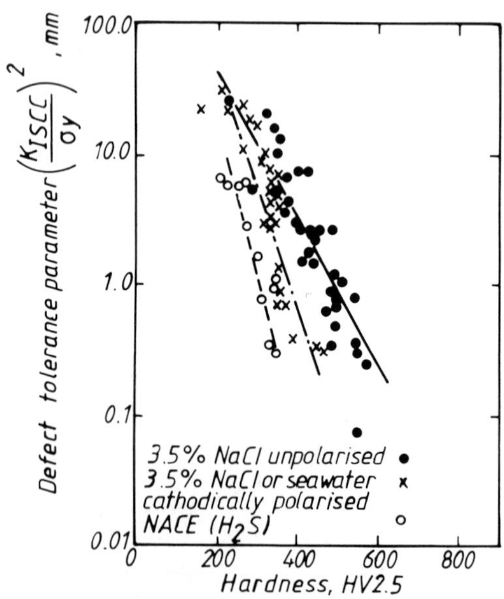

Fig. 2.4, 2 – Relationship between SCC defect tolerance parameter and hardness from precracked parent steel, HAZ and weld metal samples (excluding precipitation hardened and twinned martensite materials) (from Ref. [2.4, 9]).

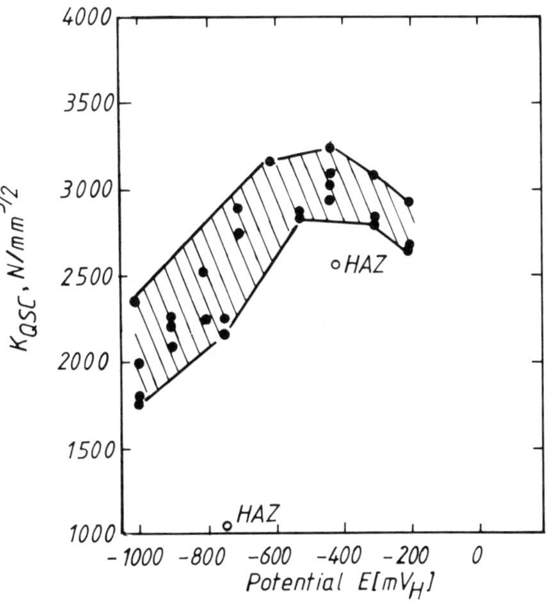

Fig. 2.4,3 – Effect of cathodic polarisation on threshold stress intensity for SCC of a fine grain low alloy steel tested in 3% NaCl or sea-water: parent steel 280 *HV*, HAZ 380 *HV* (from Ref. [2.4,18]).

higher hardness levels [2.4,9], as may arise in the heat affected zone (HAZ) of a weld, and at more negative potentials [2.4,18] (Fig. 2.4,3).

Such data do not necessarily indicate that SCC will arise in service. Again there is a paucity of information, but direct tests on welded joints in structural steels suggest [2.4,9] that, even with cathodic protection and hardened areas, static loading up to parent material yield in sea-water will not induce failure (Table 2.4,2). SCC of steel has occurred under marine conditions in service and in laboratory trials with both constant and intermittent immersion, but to the author's knowledge, only in steels of strength level above that normally employed for structural or pipeline duties. Failures in bolting and similar steels have been found with cathodic protection by zinc, although only at hardnesses above about 400 HV (Fig. 2.4,4). To some degree, this observation supports the concept of a 'critical hardness' approach during fabrication to avoid service cracking, with a hardness of 400 HV being appropriate for plain sea-water exposure. This hardness value would make no allowance for the possible action of SRB in causing H_2S nor for over-protection, and a lower level of 350 HV has been proposed [2.4,9] largely on the basis of satisfactory service of offshore structures built using welding procedures designed to hold HAZ hardness below this value to avoid fabrication hydrogen cracking. However, the situation has not been adequately defined: while available trends are encouraging in regard to structural integrity, widespread concern exists regarding the risk of SCC in offshore structures, and further testing is necessary, including levels of high cathodic polarisation as may be encountered with impressed current systems and also in conditions where H_2S is known to be present.

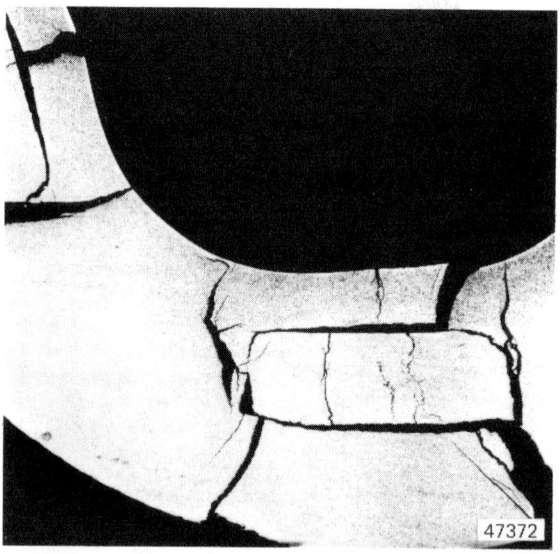

Fig. 2.4,4 – SCC in sherardised En 8 load indicating washer from service in marine conditions (× 4).

5. CYCLIC LOADING

The conjoint effect of cyclic loading and an aggressive environment in promoting corrosion fatigue is well known [2.4,1], and S–N data derived for both plain samples [2.4,19] and welded joints [2.4,19–21] show that exposure to sea-water at the free corrosion potential under conditions reproducing North Sea service causes decrease in fatigue life, with less evidence of a threshold stress for failure than found from tests in air. Given cathodic protection to about $-0.9\ V_{SCE}$ to such sample types, available data almost invariably indicate an improvement in life at low stress ranges, with, in the extreme case, fatigue lives equal to or even above those of samples in air being obtained (Fig. 2.4, 5). The improvement may be due simply to suppression of corrosion relative to the air situation, although it has been suggested that it stems also from the development of 'chalking' within propagating cracks, leading to a plugging action and effective reduction in cyclic load range [2.4, 20]. At high stress ranges, life may not be enhanced by cathodic protection, possibly because of a contribution from hydrogen embrittlement. There is no doubt that with cyclic loading, hydrogen

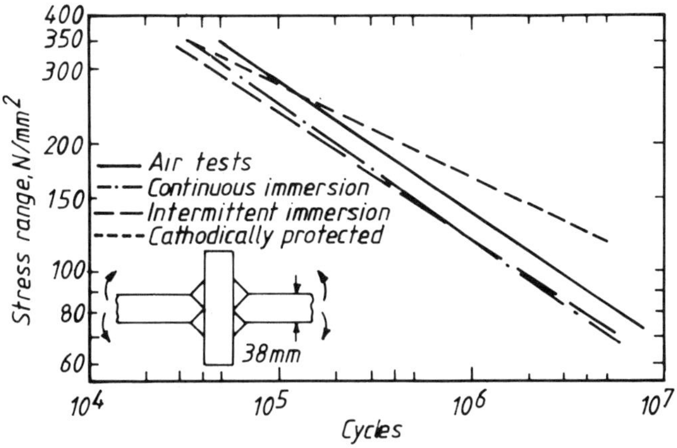

Fig. 2.4, 5 – Results of S–N tests on transverse load carrying welds in BS 4360 : Gr 50D steel at $R = 1$ in simulated sea-water with cathodic protection at -0.85 V (Ag/AgCl reference electrode) (from Ref. [2.4, 20]).

embrittlement occurs since cleavage facets, and even intergranular failure have been found on fractographic examination of cathodically polarised samples [2.4,16] (Fig. 2.4, 6). Crack propagation by intermittent brittle mechanisms may be clearly evident even at low stress (or stress intensity) ranges, but, notwithstanding this observation, cathodic protection down to about -0.8 to -0.85 V_{SCE} must be regarded as generally improving life of structural steels on

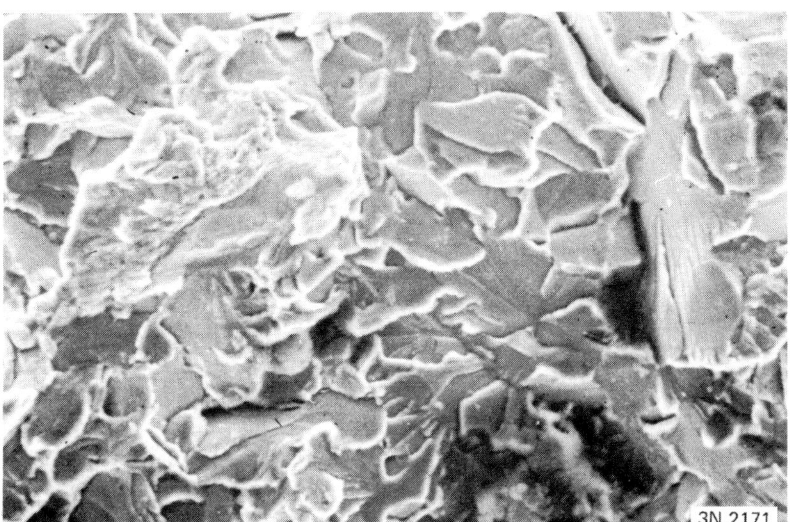

Fig. 2.4, 6 – Cleavage facets on fracture face of welded BS 3460 : Gr 50D S-N sample tested in simulated sea-water at -1.0 V_{SCE} (× 1000).

S-N testing, relative to tests at the rest potential. This, it may be noted, is the case even with welded samples having a HAZ hardness as high as 380 HV, and indeed, at about $-0.85\, V_{SCE}$, HAZ hardness is not of prime importance in determining corrosion fatigue life [2.4, 22].

More negative potentials should be regarded with some caution. At low test material hardnesses (*ca* 260 HV), polarisation even down to $-2.0\, V_{SCE}$ has been found [2.4, 23] to have no adverse effect on S-N testing, but this will not necessarily be the case with hardness of about 350 HV as can occur in weld areas: at $-1.1\, V_{SCE}$, life may be below that in air [2.4, 21], although probably still better than at the free corrosion potential.

In contrast to the beneficial effect of cathodic polarisation on S-N behaviour, data from precracked sample studies [2.4, 24-26] indicate that cathodic protection does not reduce sea-water crack growth rate to the value in air (Fig. 2.4,7). Slight polarisation to about -0.7 to $-0.8\, V_{SCE}$ gives a lower growth rate than at the rest potential, but more negative potentials worsen the situation,

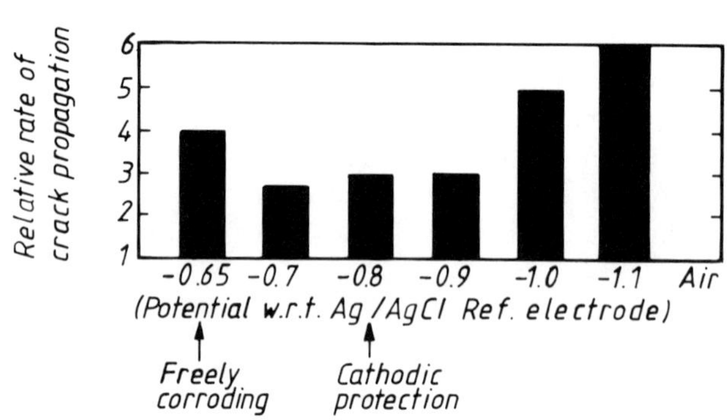

Fig. 2.4,7 – Effect of potential on fatigue crack growth rate of BS 4360 : Gr 50D steel in sea-water (from Ref. [2.4, 24]).

and considerable acceleration of cracking may be evident with over-protection to say $-1.1\, V_{SCE}$, especially at high R values (i.e. minimum : maximum stress) and with reduced frequency of loading. The reason for the anomaly between the behaviour of plain or welded samples and precracked specimens with cathodic protection is not clear. It has been suggested [2.4, 25] that, as much of the life of a plain or welded sample involves initiation and growth of small cracks, the observed enhancement of growth rate at stress intensity ranges substantially above the threshold is of secondary importance, or, alternatively in welded samples, that crack growth away from the initiation region at the weld toe relieves residual stresses, thus reducing the R value.

A further feature of crack propagation rate data under cathodic protection is that a 'discontinuity' is frequently observed in the stress intensity range/growth rate plot [2.4,25] [2.4,28], with increased growth rate at intermediate stress intensity ranges (Fig. 2.4,8). Since this has been associated with increased propensity for local crack extension by a brittle mechanism [2.4,16], it must be concluded that hydrogen embrittlement is operative in facilitating propagation. It can be remarked that the effect occurs at peak applied stress intensity levels well below those which would be regarded as necessary to cause SCC under constant load [2.4,3] [2.4,18], illustrating the adverse influence of cathodic polarisation and simultaneous dynamic strain.

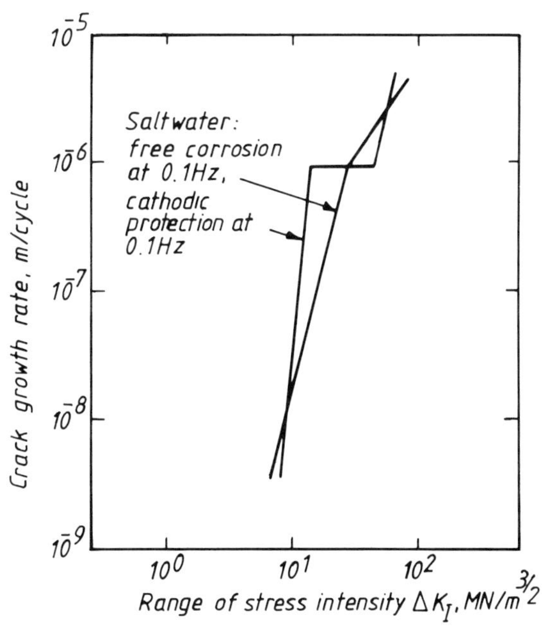

Fig. 2.4, 8 – Effect of cathodic polarisation of X65 linepipe steel on fatigue crack growth rate in 3.5% NaCl (from Refs. [2.4, 27] and [2.4, 28]).

6. PRESENT POSITION

Overall, it is clear that cathodic protection of steel structures in offshore service will induce hydrogen pick-up, and concomitant embrittlement under various loading conditions. While available data are sufficient to indicate that in most situations this will not be of practical concern, a number of questions remain, especially with regard to:

(i) the effect of gross over-protection,

(ii) the effect of local hard regions, as at welded joints,
(iii) whether SRB will exacerbate hydrogen pick-up under cathodic protection conditions.

There appear to be no failures of offshore structures which can be attributed to loss of tensile ductility *per se* as a result of cathodically-induced hydrogen pick-up under tensile loading or to the analogous problem of stress corrosion under static loading conditions. None the less, the considerable concern regarding the risk of SCC has greatly influenced fabrication procedures. Certainly, it can be remarked that defect tolerance parameters calculated from data such as those in Fig. 2.4, 2 suggest that relatively small flaws may be critical in initiating failure, even at quite low hardness levels. The concept of avoiding stress corrosion by stipulating maximum hardness values, particularly at welds, has been discussed in sections 3 and 4 above: in the author's view, current limits are conservative with normal levels of cathodic protection, but further attention to the question areas immediately above is needed. Testing should include both plain welded samples and precracked specimens, with attention to the import of microstructural variations as may be encountered in relevant materials. Moreover, on the basis that hydrogen-induced cracking under cathodic protection is promoted by dynamic strain, there would seem point in carrying out SCC tests with some cyclic load component to determine whether the risk of failure is appreciably altered.

In recent years, substantial effort has been put into the generation of realistic test data for design against corrosion fatigue of offshore steel fabrications. As a result, much information has been derived on welded joints from S–N tests under various loading and test conditions, [2.4, 29] [2.4, 30] while upper bound crack growth rate data have been acquired [2.4, 25] for a wide range of load cycle and environmental variables, including allowance for perturbations in growth rate at intermediate stress intensity ranges. Fatigue or corrosion fatigue failures have arisen in service, but to a large extent these have been because available information and design guidelines were not adequately followed. The implication is that sufficient understanding exists for service with cathodic protection as normally applied to be ensured with a fair level of certainty. However, noting that even the longest tests carried out in a laboratory have lasted for only around 10 per cent of the design life of typical offshore structures, there is no doubt that additional work is required to improve accuracy and confidence of current design procedures. Apart from the points raised above, attention to factors such as load spectrum [2.4, 30] is needed, including the effects of intermittent high or low stress hold periods on crack growth under cathodic protection. This is particularly warranted since most S–N data have been derived from fairly small samples, offering little scope for crack growth, whereas crack growth at complex joints such as node assemblies can extend over considerable distances (e.g., 0.1–1 m), when close assessment of propagation rate may be essential to predict structural life. In addition, further study on the interaction between cathodic protection and fatigue life improvement techniques such as TIG weld toe dressing is required, and on other environmental effects arising from fouling, etc. Largely because of the adverse effects on crack growth rate of cathodic polarisation to below about $-0.8\,V_{SCE}$, use of a slightly higher potential *ca* $-0.72\,V_{SCE}$ has been considered beneficial [2.4,25]. This view should be examined carefully, because such a potential may not fully suppress corrosion of the steel, especially in shielded areas: preferential attack at welds has been reported on structures with inadequate protection [2.4,27], resulting in local stress concentrations which could well enhance corrosion fatigue crack initiation at weld toes.

7. SUMMARY AND CONCLUSIONS

It has been pointed out that cathodic protection to prevent corrosion of structural or pipeline steels in sea-water will cause hydrogen pick-up. The effects of this hydrogen on material properties under static and cyclic loading have been discussed.

Although some degree of hydrogen embrittlement can be discerned, it is considered unlikely from available information that normal levels of cathodic protection will be significant in causing failure, by either stress corrosion or corrosion fatigue, even at welded joints. However, further work is necessary on the effects of over-protection, especially on weld areas of high hardness, and to examine whether or not SRB will promote increased hydrogen pick-up under cathodic protection conditions.

REFERENCES

[2.4,1] Uhlig, H. H. *Corrosion Control*. John Wiley & Sons Inc., New York, 1963.

[2.4,2] Coe, F. R. *Welding Steels Without Hydrogen Cracking*, The Welding Institute, Abingdon, 1973.

[2.4,3] Gooch, T. G. *Metal Con. & Brit. Weld. J.*, 1972, **14** (1), 29-33 and (2), 73-77.

[2.4,4] Patel, C. *et al., Proc. Second International Congress on Hydrogen in Metals*, Paris, 1977.

[2.4,5] Phalen, D. I. and Vaughan, D. A. *Corrosion*, 1968, **24** (8), 243-246.

[2.4,6] Efird, K. D. and Lee, T. S. *Corrosion*, 1979, **35** (2), 79-83.

[2.4,7] Eide, G. H. and Valland, G. *Hardness Control to Avoid Service Failures and Fabrication Cracking in Steel*, Welding Institute Seminar, Bradford, 1981.

[2.4,8] Snape, E. *Corrosion*, 1968, **24** (9), 261-282.

[2.4,9] Gooch, T. G. *Welding Institute Res. Bull.*, 1982, **23** (8), 241-246.

[2.4,10] NACE Standard MR-01-75 (1980 revision).
[2.4,11] Pussegoda, L. N. and Tyson, W. R. *Proc. Conf. Hydrogen Effects in Metals*, 1980, Wyoming, ASME, 1981.
[2.4,12] Hart, P. H. M. *Welding Institute Res. Bull.*, 1980, **21** (11), 327-333, *Proc. Third Int. Congress on Hydrogen in Metals*, Paris, 1982.
[2.4,13] Austen, I. M., Judd, W. J. and Walker, E. F. *Proc. International Conference on Steel in Marine Structures*, CEC, Paris, 1981.
[2.4,14] McIntyre, P. and Banks, T. M. *Proc. Select Seminar: European Offshore Steels Research*, UK (DoE) and CEC (C & SC), The Welding Institute, Abington, 1978.
[2.4,15] Graville, B. A., Baker, R. G. and Watkinson, F. *Brit. Weld. J.*, 1967, **14** (6), 337-342.
[2.4,16] Hinton, B. R. W. and Procter, R. P. M. *Proc. Conf. Hydrogen Effects in Metals*, 1980, Wyoming, ASME, 1981.
[2.4,17] Kasahara, K., Isowaki, T. and Idachi, H. *Proc. International Metallic Corrosion Conference*, Mainz, 1981.
[2.4,18] Kuster, F., Bohnenkamp, K. and Engell, H. J. *Proc. International Symposium Corrosion and Protection Offshore*, Centre Français de la Corrosion, Paris, 1979.
[2.4,19] Hartt, W. H. *Mat. Perf.*, 1981, **20** (11), 50-54.
[2.4,20] Booth, G. S. *Proc. Conf. Fatigue in Offshore Structural Steels*, Inst. Civ. Eng., London, 1981.
[2.4,21] Solli, O. *Sveiseteknikk*, 1978, **33** (6), 96-103.
[2.4,22] Booth, G. S. *Hardness Control to Avoid Service Failures and Fabrication Cracking in Steel*, Welding Institute Seminar, Bradford, 1981.
[2.4,23] Dvoracek, L. M. *Mat. Perf.*, 1977, **16** (9), 21-24.
[2.4,24] Scott, P. M. and Sylvester, D. R. V. *UKOSRP Report*, 3/03, Harwell, 1975.
[2.4,25] Morgan, H. G. *et al.*, *Proc. International Conference on Steel in Marine Structures*, CEC, Paris, 1981.
[2.4,26] Endo, K., Komai, K. and Kimura, K. *Bull. JSME*, 1978, **21** (161), 1565-1570.
[2.4,27] Smart, J. S. *Mat. Perf.*, 1980, **19** (5), 41-48.
[2.4,28] Vosikovsky, O. *J. Eng. Mat. & Techn.*, 1975, **97** (4), 298-304.
[2.4,29] Booth, C. S. and Vaessen, G. H. G. *Proc. International Conference on Steel in Marine Structures*, CEC, Paris, 1981.
[2.4,30] Holmes, R. and Booth, G. S. *Proc. International Conference on Steel in Marine Structures*, CEC, Paris, 1981.

2.5 COST IMPLICATION OF OFFSHORE PLATFORM STRUCTURAL PROTECTION BY COATINGS

J. DUNCAN, Salamis (Marine and Industrial) Ltd, Aberdeen, Scotland

ABSTRACT

This paper considers the economics of offshore maintenance painting and why it is necessary and cost effective. An examination is made of the elements which go to make up the cost of each square metre of coating material applied to the structure (transport, access, labour, accommodation, materials, etc.) and the justification for continuing to apply coatings throughout the life of the structure.

First, I would like to present a brief introduction to the environment with which our company is deeply involved, namely, the offshore industry, and the protection of its various large, expensive and extremely complex structures. These structures vary from small unmanned platforms such as wellhead protectors, gas platforms in the southern North Sea to the large northern oil platforms. (See Figs. 2.5,1 and 2.5,2.)

A typical large steel platform situated, for example, in the northern North Sea about a hundred miles north-east of Shetland consists of a subsea 'jacket' which is a steel or concrete tower, typically three to four hundred feet high, and a modular structure which may be another hundred feet in height. The area

Fig. 2.5,1 – Large oil platform in the northern North Sea.

Fig. 2.5, 2 – Amoco gas platform (southern sector).

which interests the maintenance painting contractor is that above the low tide level (anything below that is strictly for the divers!) and in this area the most important section, from the corrosion point of view, is the so-called 'splash zone' from the low water level to about the forty foot level (Fig. 2.5, 3). In this area cathodic protection is not available and the constant wetting with turbulent oxygenated sea-water provides ideal corrosion conditions. Studies have shown that corrosion in this area is about four or five times greater than in any other section of the platform structure and this is the area (Fig. 2.5,4) where maintenance painting activities are very often concentrated; this is, of course, also the most difficult area to tackle satisfactorily because of wave and weather problems.

The North Sea is probably as difficult an environment for maintenance painting as anywhere in the world but this has to be done and as the industry has grown so specialist companies have developed – such as our own – which are capable of managing the complex estimating, programming, logistics and operation of a major maintenance painting programme.

Planning such a programme is all important because obviously there are days when it would be impossible – if not suicidal – to work in these areas. However, with the benefit of weather statistics and past experience a balanced programme can be drawn up which, while giving priority to key structural members where corrosion might be crucial to the integrity of the structure, also include internal areas so that in the event of weather breakdown there will be other work available.

Fig. 2.5, 3 – Splash zone.

Fig. 2.5, 4 – Splash zone.

This is the background against which we have to work and it is clearly more difficult and expensive to apply coatings in these sorts of conditions than elsewhere in industry. A look at the overall cost may shed some light on just how expensive this work can be.

OVERALL COST OF A PAINTING PROGRAMME

At first glance the cost to the platform operator of putting a major offshore coatings programme into effect appears to be very large. Typically, the cost, including scaffolding, for a programme covering three or four platforms, could be over five million pounds in a single year. Unfortunately, much of this does not go into the contractor's bank account as the associated support facilities provided by the platform operator absorb the lion's share. A breakdown of the logistics support required to keep a single coatings applicator makes interesting reading and this may be seen in Table 2.5,1.

Table 2.5,1
Coatings applicator

Helicopter	Supervision
Supply boat	Scaffolding
Accommodation	Management onshore
Standby vessel	Back-up labour
Cranes	Materials supply
Inspection	Vocation training
Safety	Safety training
Catering	
Offshore support	
Specification	

If a typical large offshore painting and scaffolding contract was valued at around £1 million to the contractor then it would probably cost the oil company four or five million pounds to effect that programme fully, i.e., taking all the necessary support items into account. For example, an accommodation barge may cost £30 000 per day to accommodate 250 men – so immediately each man costs the operator £120 per day (or £10 for each hour employed) and this is just for bed space.

Various estimates have been made of the true cost of employing a man offshore and, depending on the amount of 'back up' each man requires, these estimates vary from £40 to £50 per man-hour (say £45 per man-hour). Of this the painting contractor on an hourly rate contract would hope to receive about £10 to cover all his labour costs and overheads. The details may be seen in Table 2.5,2.

Table 2.5, 2
Man-hours

Productive	1.0	(actual coating)
Back-up services	0.8	
Downtime	0.5	
Total	2.3	man-hours

Each man-hour costs £45

For example, splash zone
1 square metre requires 7 man-hours, so real labour cost is *£315 per sq. metre.*

From our experience of offshore contracting we have learned that for every directly productive man-hour there are about 1.3 non-productive man-hours, so taking this ratio of 2.3 actual to productive man-hours we can see that each productive man-hour is costing £45 × 2.3, i.e. £103. When it is realised that it can take up to 7 man-hours to blast and coat one square metre at the tidal zone including all the non-productive hours then one can see that the total cost to the operator can be over £300 for that square metre – without materials. Actually, it would be marginally cheaper to paper that square metre in £1 notes, two layers thick – although this provides no corrosion protection! Other areas above the splash zone would cost the operator one-third of these rates as they are less susceptible to disruption by weather and require less scaffolding.

NON-PRODUCTIVE TIME AND DOWNTIME

Non-productive time is caused by various factors. In any offshore painting contract good supervision is essential but a supervisor is essentially non-productive. There will also be chargehands reporting to the supervisor who will be only partly productive, and, the necessary safety men and back-up labourers who, although essential, do not actually produce coated steel.

Downtime is time lost through various reasons, the main ones of which are:

(1) Weather disruption which can account for 30 per cent of lost time and much more when working in the splash zone.
(2) Time in transit (to or from the shore or shuttling from the accommodation barge).
(3) Platform operations, i.e., boat drills, obtaining work permits, crane operations, etc.
(4) The unforeseen, e.g., mechanical breakdown of equipment, lack of supplies due to storm delays, etc.

It is important to note that, compared with the cost of putting labour on the job, material costs are negligible. Thus, typical coating materials cost about £5–£6 per square metre and therefore the very best material should always be chosen so as to delay as long as possible the need to return to the completed area.

Naturally, at these sorts of cost levels it is important to obtain a high level of efficiency and a well motivated work-force. However, the labour force is located on isolated structures which are not less than 150 miles, and perhaps as much as four hundred miles, from headquarters. Of necessity, management must be from this distant office as there is no room on the platform for permanent personnel who are not directly engaged in producing completed work. Life on a platform for a painter is generally fairly boring after the first few days of his two-week period and the best motivation is good pay.

The necessity is, therefore, to employ hardworking, conscientious workmen who can work with little supervision since it is normal for them to operate in small squads at various locations on the platform. However, a good offshore supervisor is essential to co-ordinate the programme efficiently and to reduce expensive lost time as much as possible.

Since, as we have seen, it costs so much to paint a square metre offshore, then if an efficient squad can increase output by as little as five per cent the savings obtained in completing a 100 days programme in 95 days will be dramatic. Thus, the cost of paying a little more for good labour is insignificant compared with the potential savings. Fortunately, the oil companies are aware of this and a good track record is now essential in qualifying for the opportunity to tender for a major contract.

EQUIPMENT

The previous remarks apply equally well to equipment and five per cent lost time through equipment failure is a loss as much as that through labour inefficiency. A necessary stock of spares is essential as replacement of faulty equipment can take several days by supply boat. Personnel must be trained in the basics of equipment maintenance. New equipment is constantly being developed and tested offshore in the hope of improving output but sometimes the complexity of the equipment means that it is not easy to maintain in such isolated locations as offshore platforms. The ideal piece of equipment for offshore is simple, robust and with the minimum of moving parts. A shovel is just perfect!

COATING MATERIALS

With offshore man-hours so expensive, anything which can make these more productive is valuable and so coating materials are being developed which will extend the time when it is possible to work – since every man-hour saved is worth £45 to the operator. In past years work would have stopped because of damp steel or high humidity. From the point of view of economics this approach is much more important than that of the coating manufacturer cutting his prices to the bone to save a few pence on the material cost. As already described, the cost of coating material is almost negligible compared with the cost of application and so if the material has significant built-in time-saving advantages then these will generally far outweigh the extra cost. The ideal coating material is one which can be successfully applied in a single coat over untreated and wet surfaces and which will last for the life of the structure, but it does not exist!

However, in order to reduce the huge labour costs materials of the future must have some of these characteristics and it seems that manufacturers have already gone a long way towards achieving some of these desirable properties. Moisture-tolerant materials have already been mentioned and some other major advances which could have significant savings in labour costs include the following.

(1) Coated blasting grit, which deposits a layer of primer as the blasting process continues and so allows work to proceed in damp weather.
(2) High build topcoats which cure rapidly at low temperatures allowing blasting operations to continue in adjacent areas with minimum lost time.
(3) Materials which can be applied to the full required thickness in a single coat.

SCAFFOLDING

The access system is probably the most expensive part of a painting contract system. Conventional scaffolding, as shown in Fig. 2.5, 5 is widely used and is generally very effective but in bad weather it is very vulnerable.

A conventional scaffold system can be destroyed overnight (Fig. 2.5, 6) and a major removal exercise will be necessary to replace it.

Developments in this field include a new mobile cradle which has been designed to travel along angled (Fig. 2.5,7) and horizontal braces (Fig. 2.5, 8) and which can be assembled and rigged in a few hours. Since it is self-powered it can be quickly winched up to the higher levels when bad weather threatens. Although this will not eliminate conventional scaffolding altogether it may reduce the need to build such large fixed structures in these vulnerable areas and consequently reduce the number of highly expensive man-hours involved in their construction.

SAFETY

This is a subject which is not often considered as a cost item in a painting contract but offshore safety is expensive, although not easily quantified; the benefits are seldom tangible – which is perhaps just as well. Safety for the

Fig. 2.5, 5 – Conventional scaffolding.

Fig. 2.5, 6 – Storm damaged scaffolding.

Fig. 2.5,7 –
Mobile scaffolding cradle.

Fig. 2.5, 8 –
Mobile scaffolding cradle.

contractor usually starts in the cost of sending all his offshore personnel to the offshore survival course and the fire-fighting course. Furthermore, all his offshore equipment must be fitted to stringent safety standards. I shall not detail these here as space forbids but suffice to say that a compressor fitted to full offshore safety specifications costs twice as much as the same compressor on a building site. These costs and the cost of boat drills and fire drills resulting from lost man-hours are all part of the painting cost equation which must be taken into account. However, the cost of *not* maintaining an adequate safety policy is too high as the bill may be in lives rather than in pounds.

One can thus say that painting offshore is difficult and expensive. The economics can be improved only by putting effort into getting labour at the workface for longer periods, i.e., by reducing downtime. This can be achieved by:

(1) Motivation and training.
(2) More reliable equipment.
(3) Coating materials which are more tolerant of marginal conditions of humidity and temperature and which require fewer coats.
(4) More rapid access to difficult areas.

One might ask 'Why bother with painting?' All platforms are built with a reasonable corrosion allowance for the design life which might be 25 years. However, there will inevitably be damage to the protective coating over the years and reliance on the corrosion allowance alone is not generally acceptable because there will be variations in rates of corrosion from those which have been predicted at the design stage. Additionally, and perhaps more importantly, there seems to be a tendency recently to load existing platforms with extra weight in the shape of new modules (Fig. 2.5,9) for various requirements. These may be for gas conservation measures (since recent legislation has discouraged flaring off surplus gas) of for extra accommodation, a heavier helideck, or other purposes. In some of these cases extra structural strength necessary to support the extra weight has been found by utilising some of the corrosion allowance built into the structure. It thus becomes even more important to ensure that the diminished corrosion allowance should be protected so as to allow the platform to complete its design life and, above water, protective coatings are the only economic method at present available.

Typically, the corrosion allowance in the splash zone of an offshore platform will be about half an inch, and as the annual rate of corrosion is generally reckoned to be about one mil (or 1/25 in) per year, then the corrosion allowance will be used up in about half the expected 25 years' life of the platform. The corrosion allowance is there to be used, and ideally, a four-year cycle after painting would give two years of protection followed by two years of corrosion followed by another repaint. This would mean that no more painting would be required than the minimum to ensure that the platform remained with its structural strength unimpaired for its planned life.

Fig. 2.5, 9 – Module interior.

However, the increasing value of crude oil and better methods of recovery mean that it may well be necessary and economical to increase the working life of the platform by several years. As these structures can cost as much as £500 million to replace at current values then a painting programme costing even one million pounds every two years or so is much cheaper than replacing the platform or carrying out major reconstruction work. This is basically the justification for protective coating application.

2.6 COST EFFECTIVE USE OF STAINLESS STEEL IN TOPSIDE MODULE CLADDING AND STRUCTURAL APPLICATIONS

R. A. E. HOOPER, British Steel Corporation Sheffield Laboratories, Swindon House, Moorgate, Rotherham

ABSTRACT

Topside modules on offshore platforms are conventionally clad with mild steel, protected from corrosion by galvanising and/or painting. The North Sea marine environment is severe and causes coating failure in a few years, which necessitates inconvenient and very expensive maintenance painting.

The use of light gauge stainless steel cladding eliminates maintenance and is cost effective over the lifetime of most platforms. The cladding has full fire resistance up to, and including H120 (120 minutes in a hydrocarbon fire) and the weight saving leads to more cost savings in the support structure.

The modular, pre-insulated system has now been proven and is being specified and used in an increasing number of North Sea platforms.

1. INTRODUCTION

Most of the topside equipment of an offshore production platform (gas or oil) is made from carbon steel that is protected in some manner from the environment. The offshore atmosphere is very corrosive, consisting at best, of salt-laden air and, at worst, of salt-laden air containing wind blown salt spray or actual salt spray coming directly from a heavy sea. In practice, paint coatings have a limited life in this situation although structural members are made from sufficiently heavy sections to allow for a degree of corrosion before maintenance painting can be carried out.

The topside of a platform is densely filled with modules of various types, for living and working, and most are characterised by a steel shell structure of walls, floors and ceilings (Fig. 2.6,1). Conventionally, these have been fabricated from carbon steel plate protected from corrosion by hot dip galvanising with zinc and painting onshore prior to float-out. Whilst this technique gives a good life for land-based structures, in the North Sea there can be significant breakdown of protective coatings within 3–8 years. The early failure of these coatings, often by mechanical damage, leads to rapid corrosion of the substrate steel giving less protection of adjacent areas by adhesion failure, Consequently, maintenance painting is essential to prevent the loss of the structural and thermal integrity of the modules. Offshore maintenance painting is difficult, due to the severe weather conditions, which generally prevail, and the quality of workmanship usually is considerably inferior to the original painting onshore. This leads to an even shorter coating life before the next session of maintenance painting and the situation can progressively go from bad to worse as rectification becomes more difficult.

Fig. 2.6,1 – Offshore oil platform.

Furthermore, offshore maintenance is very expensive. All materials, scaffolding, blasting equipment, paint and consumables have to be transported and stored offshore. Additionally, painters and inspectors must be flown out and beds need to be found for them whilst work is in progress. Every platform seems to have a bed shortage and each maintenance bed means one less production man and consequently platform production and efficiency falls. The cost of lost production can be enormous and frequently incalculable but the cost of having a maintenance man and his equipment offshore is said to be about £500 per day, excluding consumables, whether or not the climatic conditions are suitable for him to work!

Clearly, therefore, there is a need for a cladding system with lower or no maintenance requirements and stainless steels are a feasible alternative.

2. STAINLESS STEELS

2.1 Composition

Stainless steels differ from mild and low alloy steels and are characterised by the fact that they contain at least 10 per cent chromium. This results in the formation on the steel surface of a chromium rich oxide film. This film is extremely thin and hence invisible, but it is inert, very adherent and extremely protective; it is referred to as the passive film.

Unlike applied coatings, the passive film has the unique property of self-repair, such that it reforms instantaneously in the presence of air, or other oxidising media, if it is mechanically damaged or removed. The passive film confers on to the stainless steel a resistance to corrosion in a wide range of environments but the degree of protection is dependent on the chemical composition of the oxide film, which in turn is related to the composition of the steel.

The chemical composition of stainless steels defines the structure of both the steel and the passive film. The principal alloying element is chromium; carbon is generally restricted to levels of <0.06 per cent because chromium reacts with carbon to form carbides which can reduce the corrosion resistance in some circumstances, Table 2.6,1.

Table 2.6,1
Typical chemical composition and RT mechanical properties of some standard stainless steels

Specification		Structure	Composition, wt%				Mechanical properties			
BS	AISI		C	Cr	Ni	Others	0.2% PS (N/mm^2)	TS (N/mm^2)	% (El)	Izod (J)
409S17	409	α	0.05	12	–	0.4 Ti	295	480	33	25
430S15	430	α	0.05	17	–	–	340	540	26	25
304S16	304	γ	0.05	18	10	–	240	600	55	150
316S16	316	γ	0.05	17	12	2.5 Mo	260	590	51	165

In commercial alloys, the lowest chromium content employed is about 12 per cent (Type AISI 409), but significant improvements to the oxide film stability are afforded by increased chromium additions. Most stainless steels contain about 18 per cent chromium because above this level further chromium additions provide only small additional benefits. Additions of about 9 per cent nickel to the Fe–18% Cr base (Type 304) have little effect on the passive film but do significantly raise the electrochemical nobility of the steel, and also provide considerable improvements to the mechanical properties. Better corrosion resistance is obtained by small additions of molybdenum to the Type 304 composition, as in the commercial Type 316 (2.5% Mo).

2.2 Structure and properties

The commercial 12 to 18% Cr stainless steels possess a fully ferritic structure which provides good tensile strength (450–600 N/mm^2) and a moderate ductility (20–30% elongation). The addition of 8–10 per cent nickel to the composition, e.g. Fe + 18% Cr, 10% Ni (Type 304) steel, produces a fully austenitic structure which possesses enhanced tensile properties and considerably increased ductility (50–65% elongation). This steel also possess a high work hardening rate which is an advantage in providing increased rigidity in pressed panels. The addition of small amounts of molybdenum to the Type 304 composition (i.e. 2.5% Mo in Type 316) does not affect the austenitic structure and has no detrimental effect on mechanical properties, Table 2.6.1.

2.3 Corrosion resistance

The commercial ferritic stainless steels possess considerably improved corrosion resistance compared to mild steel due to the passive chromium-rich oxide surface

Fig. 2.6, 2 – Pitting corrosion.

film, and the degree of protection afforded by the film increases with increasing chromium content up to about 17% Cr (Type 430). Nevertheless, the destruction of the protective film by oxidation, reduction or local penetration of the type typical of chloride attack, can occur in many chemical environments. The commercial austenitic Type 304 composition possesses the additional protection afforded by the noble effect of nickel, which at levels above 5 per cent provides much improved resistance to corrosion. However, despite the good general corrosion resistance of Type 304, it remains prone to chloride penetration of the oxide film which can lead to pitting corrosion, Fig. 2.6, 2. The addition of a few per cent of molybdenum to the Type 304 composition, as in the commercial Type 316 alloy, produces improvements to corrosion resistance in certain dilute acid environments. More importantly, however, molybdenum promotes passivation of the steel under conditions of localised film penetration typical of chloride attack, and it is this property which substantially increases the resistance of Type 316 to chloride induced pitting corrosion, Fig. 2.6, 3.

3 CORROSION

3.1 Atmospheric corrosion

When thinking of atmospheric corrosion resistance, it is often in the context of the building industry, where stainless steel is used for its aesthetic attractiveness in columns, rails, roofs, wall cladding, etc. In this sphere, the conventional recommendations are Type 430/304 for internal use, Type 304 for exterior use in rural and urban environments and Type 316 for urban/industrial and marine situations.

In offshore installations the structures must be strong, sound and reliable rather than attractive and elegant; appearance is less important. It is the actual rate of corrosion that matters in the fiercely aggressive marine climate. Mild steel has a higher metal wastage rate and therefore needs treating regularly with protective coatings, which are very expensive to apply and maintain. Stainless steel offers great cost benefits because it needs no regular, time-consuming and costly maintenance.

3.2 General corrosion

If the surface oxide film is completely destroyed, then all of the metal surface will be exposed to attack and if the metal is homogeneous, it will be attacked in a uniform manner. In this case, it is possible to quote to the engineer a corrosion rate value from which he can estimate the metal thickness required to give the desired life. For example, general surface corrosion is a feature normally exhibited by stainless steels when attacked by acid media and by mild steel in nearly all environments.

In offshore applications, this type of situation is unusual with respect to stainless steel and the localised corrosion mechanisms explained below need more important consideration.

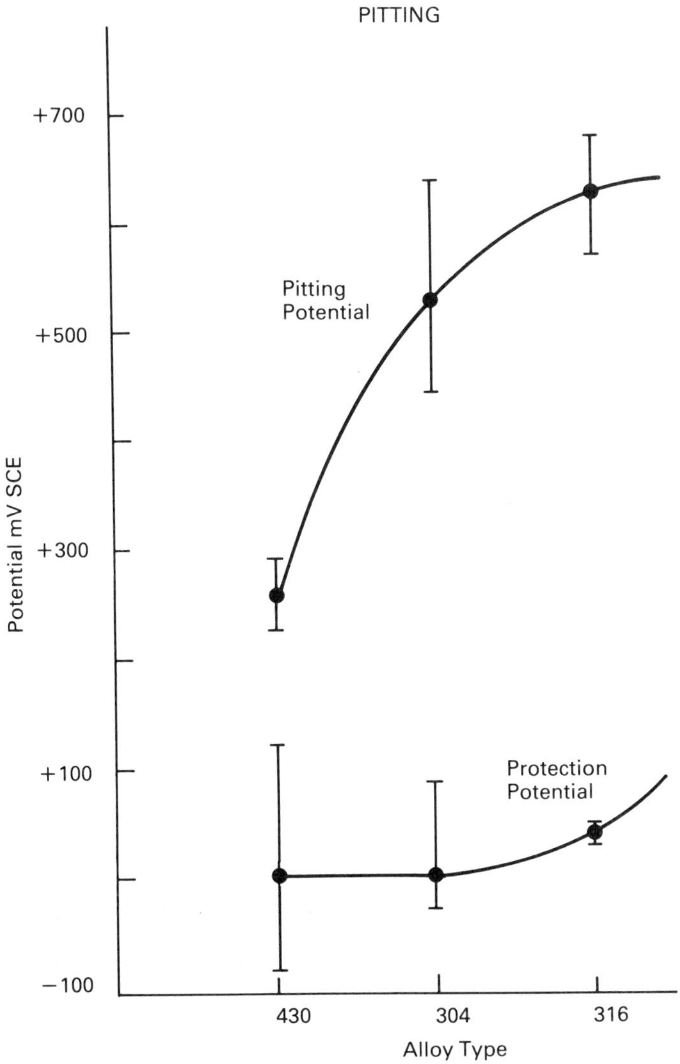

Fig. 2.6, 3 – Pitting and protection potentials of stainless steels.

3.3 Pitting corrosion

Pitting corrosion is a result of local disruption of the oxide film due occasionally to material imperfections, for example sulphides, but more usually to an environment that is sufficiently aggressive to cause local break-down of the passive film without general dissolution. This form of attack cannot be adequately described in terms of loss in weight or average loss in thickness, because penetration by pitting can occur when such values remain quite small. Pitting corrosion may be prevented on many occasions by using molybdenum-bearing austenitic stainless steels because molybdenum is effective in considerably strengthening the oxide film, thereby enabling it to remain unattacked in environments of quite aggressive natures.

3.4 Crevice corrosion

Crevice corrosion is a special form of pitting corrosion where the passive film is disrupted by oxygen starvation caused by the shielding of a crevice geometry. Oxygen is important in maintaining passivity and failure to maintain this condition over the entire surface can lead to preferential corrosion at those sites starved of oxygen. Such conditions can exist for example beneath a washer in a bolted construction, beneath fouling organisms in the sea, for example barnacles, or even at a crevice formed between pieces of the same material, Fig. 2.6,4. Whatever the situation, a crevice will only present a corrosion hazard if it is wide enough to permit entry of a liquid and sufficiently narrow to maintain a stagnant zone. For these reasons, crevice corrosion usually occurs at openings a few tens of microns or less in width and it rarely occurs within grooves that are several millimetres wide. In consequence, it is possible for some gaps, which might be defined as crevices, to be relatively safe but a precise decision is not really possible without experience of the situation involved and, therefore, the general tendency is to recommend their elimination. Crevice corrosion can be minimised by using a molybdenum bearing stainless steel, e.g. Type 316 (containing 2½% Mo).

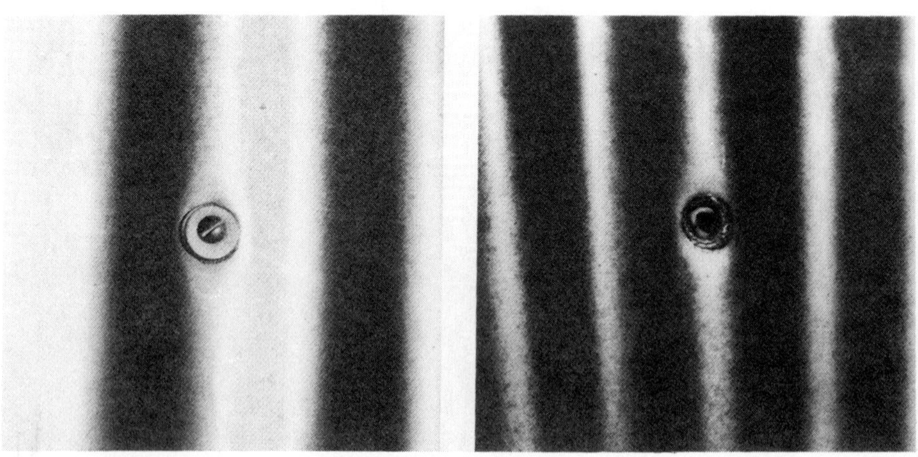

Fig. 2.6,4 – Crevice corrosion of Type 304 stainless steel under a washer after 34 years of marine exposure (courtesy of INCO).

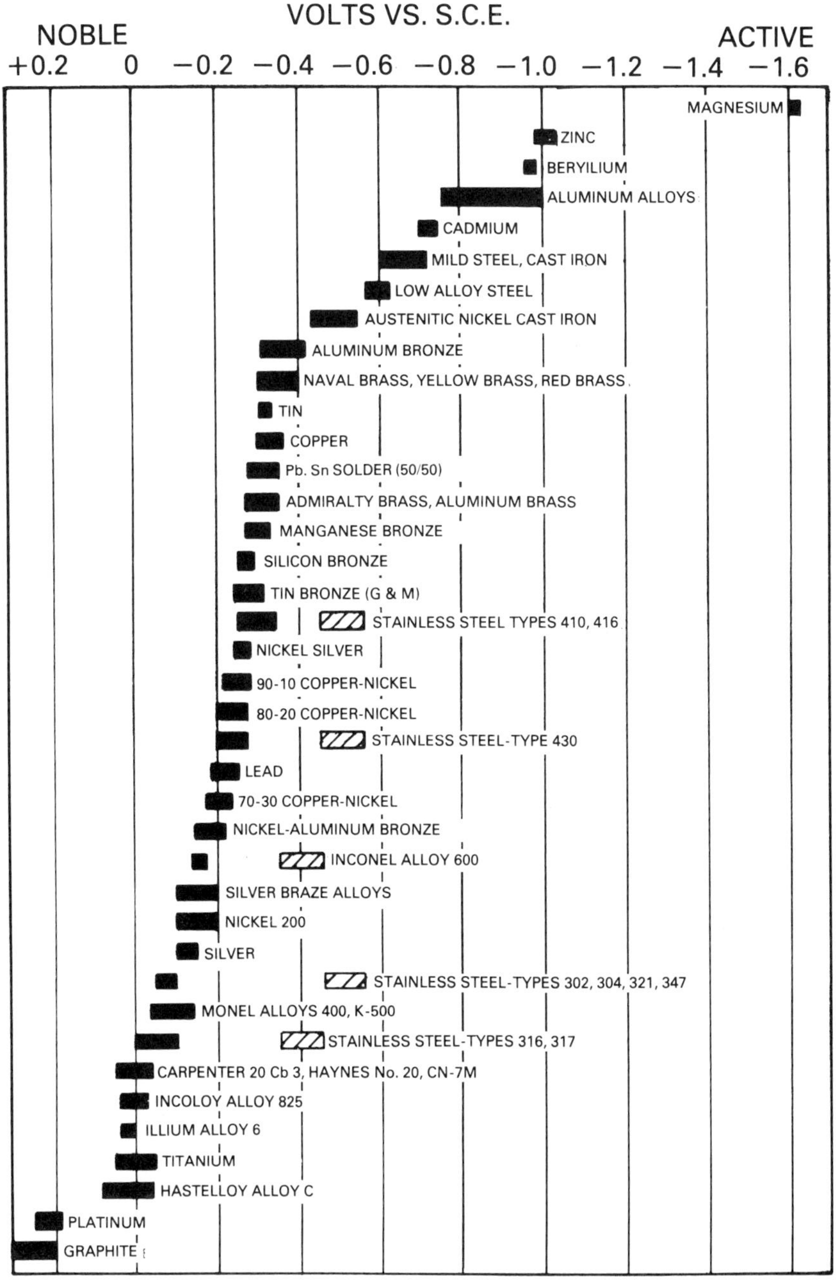

Fig. 2.6,5 – Galvanic series in sea-water (after Sedriks).

3.5 Galvanic corrosion

Galvanic corrosion occurs when dissimilar metals are placed in electrical contact in a corrosive solution and this results in the corrosion rate of each metal being altered. Very long periods of exposure to highly conducting, aggressive solutions are usually needed to cause significant galvanic corrosion and it is therefore quite a rare corrosion mechanism except under conditions of total immersion.

The further apart that the metals are in the galvanic series, the greater the effect. The corrosion rate of the more noble metal is reduced and the less noble is accelerated (Fig. 2.6,5). If zinc and mild steel are exposed together, for example at the edge of galvanised sheet, the zinc will dissolve preferentially and suffer by accelerated corrosion, whilst the mild steel will be protected. In the case of galvanised sheet, this form of galvanic corrosion is intentionally used to protect the steel. There is also an area effect in that the relative areas of each metal exposed also control the rate of corrosion. If the less noble metal has a smaller surface area than the more noble metal, it will corrode even more rapidly.

Stainless steels are the most noble of the common engineering and construction metals and, therefore, rarely suffer galvanic corrosion but their effect on less corrosion resistant metals that might be coupled to them needs consideration. When painted mild steel is joined to stainless steel it is good practice to paint over the joint and to cover a few centimetres of the stainless steel to prevent the possiblity of galvanic corrosion of the mild steel.

3.6 Intergranular corrosion

Intergranular corrosion, or weld decay, was associated with welded heavy sections of high carbon stainless steel, particularly if a post weld heat treatment was omitted.

In high carbon (>0.06%) austenitic stainless steels, chromium-rich carbides precipitate at the grain boundaries in the parts of the weld heat affected zone that experience exposure to a temperature range of 500 to 750 °C. These precipitates draw chromium from the adjacent matrix leaving a region susceptible to corrosion attack. This attack is, however, entirely preventable by the choice of the correct grade of stainless steel. Historically, titanium, and occasionally niobium, stabilisation has been used as these elements combine with carbon to form carbonitrides on cooling after solution treatment. Consequently, there is insufficient carbon in the matrix to form chromium-rich carbides during welding. The alternative method of preventing intergranular corrosion is to use a low carbon steel (<0.03%) but in the past the low carbon grades have been difficult to manufacture. However, the advent of modern AOD practice has enabled carbon to be reduced in an efficient and consistent manner. Low carbon grades are now slightly cheaper than the stabilised grades and are being increasingly preferred.

When welding thin (<3 mm) stainless steel by modern electric techniques

(TIG, MIG and MMA) the heat input is insufficient to cause chromium-rich carbide precipitation in standard, austenitic stainless steels of typically 0.05% C. Therefore, there is no need to specify low carbon (<0.03%) or stabilised grades to prevent weld decay in thin sheet.

3.7 Corrosion fatigue

The most important factor in corrosion fatigue is the corrosion resistance and not the tensile strength. No heat treatment which raises tensile strength without increasing corrosion resistance is likely to be beneficial but a heat treatment which impairs corrosion resistance will impair corrosion fatigue resistance.

The first step in corrosion fatigue is believed to be the occurrence of a pit which ultimately intensifies the actual stress. Therefore, the higher alloy stainless steels with good pitting corrosion resistance also have the best corrosion fatigue resistance.

3.8 Stress corrosion

Stress corrosion occurs in almost all alloys under specific circumstances. In process plants the conditions that most commonly need to be guarded against when using stressed austenitic stainless steels are hot, concentrated chloride solutions.

Classical, transgranular, chloride induced, stress corrosion cracking is very rare at temperatures below 50 °C, but it needs consideration for hot pressure vessels, hot storage tanks and heat exchangers. A few hunderd parts per million of chloride can be tolerated for many years provided that the operating temperature is not too high, but design is important to prevent accidental concentration.

The common sites of concentration are the hot water line, heat exchanger surfaces and crevices. When an aggressive environment is formed by a concentration at such locations, an improved resistance to stress corrosion can be obtained by using a molybdenum bearing austenitic stainless steel. Molybdenum itself does not improve stress corrosion resistance but helps to prevent the localised attack which generates the aggressive conditions.

4. MARINE ATMOSPHERIC EXPOSURE DATA

4.1 British data

Stainless steel panels, encompassing a range of chemical compositions, have been continuously exposed over a 17-year period in both the sheltered and unsheltered situations in a marine atmosphere 15 m from the sea at Rye, Sussex. At 6-monthly intervals, half of each panel was washed with soap and water to remove debris and encrustations, and a visual examination of both halves of the panels was carried out. This indicated superficial rust staining to varying degrees, dependent primarily on the chromium and molybdenum content of the material, Fig. 2.6,6. This was heavier on the low alloy, ferritic alloys, but reduced to

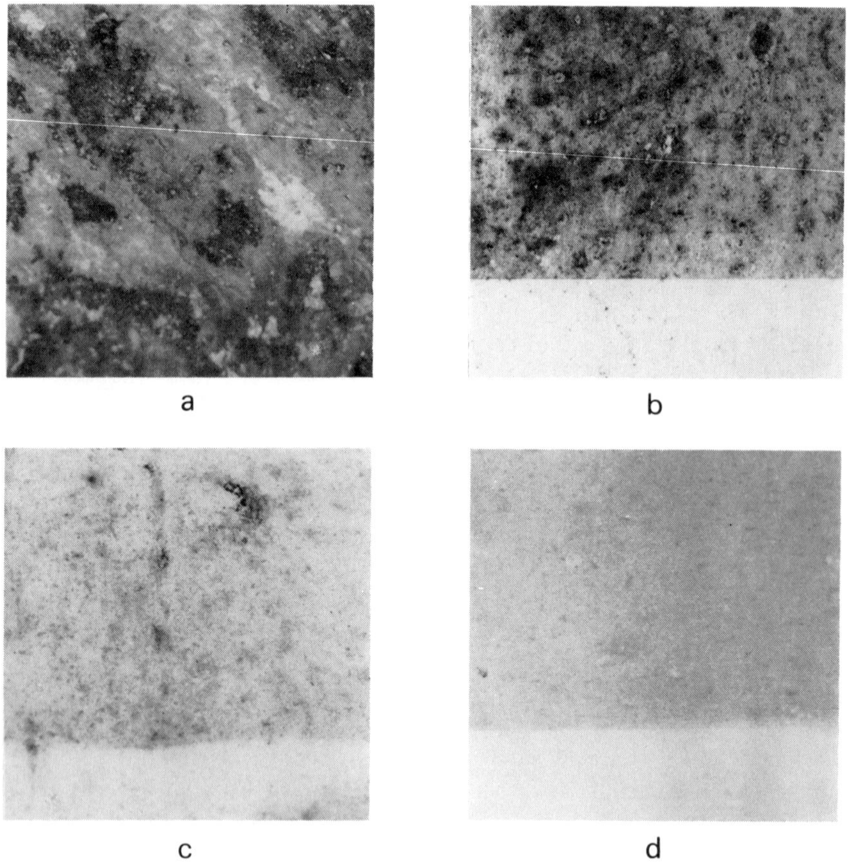

Fig. 2.6,6 – Atmospheric corrosion of sheltered steels at a marine site: (a) Type 409 after 1 year; (b) Type 430 after 15 years; (c) Type 304 after 15 years; (d) Type 316 after 15 years.

minimal levels in the higher alloyed austenitic alloys in the order 409, 430, 304, 316. Most of the deterioration in terms of rusting and pitting occurred within the first few years of exposure after which further corrosion took place at very much lower rates. In all cases, the degree of corrosion was greatest in the sheltered panels due to the retention of deposits and the concentration of chloride salts. Washing of the unsheltered panels, either manually or by natural rainfall, markedly reduced surface contamination, and resulted in lower levels of corrosion.

Cleaning of the Type 304 and 316 panels with a non-abrasive cleaning compound easily removed all of the superficial rust staining and restored a bright finish. Examination of the cleaned panels indicated isolated and shallow pitting, generally only visible under close examination, and this was at a minimum with the Type 316 material.

4.2 American data

The atmospheric exposure tests were carried out over periods ranging from 23 to 40 years at Kure Beach, Wilmington, North Carolina at distances of 25 and 250 m from the Atlantic Ocean. The conditions pertaining to this site differed from the British site primarily in being hotter and having a lower rainfall, both of which combine to provide a more aggressive environment. Despite this, and the longer exposure periods, the Type 304 and 316 panels exhibited corrosion levels not significantly greater than the British test panels, and the rust staining and pitting were confined to a superficial level.

5. WELDABILITY

The austenitic stainless steels are readily weldable and these alloys retain an austenitic face-centred cubic structure down to sub-zero temperatures and do not exhibit excessive grain coarsening at elevated temperatures. These two factors are largely responsible for the excellent weldability of the materials since there is no danger of the materials hardening due to the weld thermal cycle and the joints remain ductile and tough in the as-welded condition, Fig. 2.6,7.

During arc welding there is an appreciable temperature rise at the joint. The peak temperature reached and the time spent at elevated temperatures are determined by the heat input and the thickness of the materials being joined.

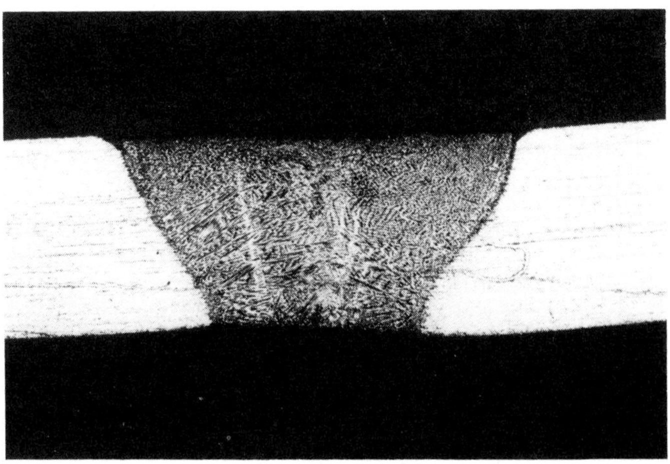

Fig. 2.6,7 – Autogenous TIG weld in austenitic stainless steel.

Most modern welding processes produce essentially low heat inputs and are unlikely to cause any reduction in corrosion resistance except in extreme cases such as the multi-pass welding of thick sections.

5.1 Electrodes

Thin (<3 mm) austenitic stainlesss steel is frequently welded by the autogenous TIG technique (no filler metal) if the fit up is good. However, filler wire and consumable electrodes are available, which match the base material both in strength and corrosion resistance. The composition is usually slightly different to the parent stainless steel so that, on solidification, the weld metal contains a small proportion of delta ferrite. This is to ensure that the welds are free of cracking. The amount of delta ferrite can vary with the welding process, material and joint configuration but usually lies between 2 per cent and 10 per cent.

Filler wires, for use with the MIG and TIG processes, require no special handling precautions other than keeping them free from oil, dirt or grease, which could introduce carbon into the weld. To prevent porosity in the weld, manual metal arc electrodes should be stored under proper conditions to avoid moisture pick-up in the coatings. Typical storage conditions should be a minimum temperature of 15 °C and a relative humidity below 50 per cent.

5.2 Design

The design of joints in austenitic steels is generally straightforward but there are a number of points which should be borne in mind. For example, if small tight crevices are present in welds in contact with liquids, crevice corrosion can occur. Sensible design, using fully penetrated welds wherever possible and taking precautions to avoid weld metal solidification cracking, will help to prevent this occurrence. Perhaps one of the biggest problems when welding austenitic stainless steels is the control of distortion. Austenitic steels have a coefficient of thermal expansion some 50 per cent greater than carbon steel and a thermal conductivity only one-third of carbon steel. This means that temperature gradients can build up rapidly causing unequal expansions and distortion. This can be minimised by a planned approach to the welding of each joint. Low heat input, balanced welding, symmetrical joints, good fit up and well designed jigs are all desirable.

Preheating of austenitic steels is not recommended since the aim is to prevent excessive build up of heat.

5.3 Dissimilar metal welds

There is frequently a need to weld stainless steels to constructional steels and it is fortunate that austenitic stainless steels are readily weldable to mild or low alloy steels. When selecting the most suitable procedure, the effect of the dilution of the weld metal by the base materials must always be considered. In normal joints the weld metal can be diluted with between 20–50 per cent of the base materials depending on the type of joint and the disposition of the individual weld beads. For example, a root run in a prepared butt joint would be heavily diluted whereas the capping runs would retain almost the undiluted weld metal composition.

The actual choice of welding electrode is very important since it must be capable of developing the correct physical properties over a wide range of dilutions. Mild or low alloy steel electrodes are never used since the pick-up of chromium and nickel from the stainless steel base material would lead to a hardenable weld metal composition resulting in a hard and crack prone weld. The accepted procedure is to use an over-alloyed austenitic electrode for the dissimilar metal joint. This type of electrode is designed to produce a weld metal of approximately 18% Cr 8% Ni after normal dilution, which ensures adequate mechanical properties and corrosion resistance. In addition, the composition is balanced so as to retain a small amount of delta-ferrite in the microstructure to ensure freedom from cracking during welding. Typical electrodes of this type have compositions approximating to: 23% Cr 12% Ni; 23% Cr 12% Ni, 2.5% Mo; and 20% Cr 10%Ni 3% Mo.

Fig. 2.6, 8 – Stainless steel module wall section after fire testing.

6. FIRE RESISTANCE

Fire resistance is defined by referring to the SOLAS (safety of life at sea) specifications, which are themselves drawn from IMCO (Inter-Governmental Maritime Consultative Organisation) International Conference First Act 1974. Panels manufactured to these specifications are proven by the fire resistance tests described in BS 476: Part 8, 1972 in order to gain approval from the relevant certifying authority, e.g. Lloyd, DNV. The stainless steel clad panels pass the A60 test by a significant margin, that is to say it easily withstands a fire, as defined in BS 476 for 60 minutes.

A more severe test is the resistance to hydrocarbon fire because in this case the rate of rise of temperature is greater (one side of the panel reaches over 1100 °C in 30 minutes). This is not covered by BS 476; the heating rate curve was developed by the Norwegian Petroleum Directorate and the Mobil Oil Co. Ltd. The terminology is similar, for example A120H refers to a steel (A) clad panel resisting a hydrocarbon (H) fire for 120 minutes. A120H is frequently referred to as H120. The 1.6 mm thick stainless steel clad insulated panel withstands the H120 test without difficulty as shown in Fig. 2.6, 8.

7. MATERIAL CHOICE

For offshore accommodation modules, the principal form of corrosion to which cladding panels will be subjected is chloride induced attack resulting from the marine atmosphere. Mild steel, under such conditions, suffers general surface corrosion leading to perforation of the panel section in spite of protective coatings. Stainless steels, in either the painted or unpainted condition, can be expected to offer substantially longer life and very low maintenance costs.

Marine atmosphere exposure tests in both the UK and USA for periods of 15 to 40 years have indicated only superficial rust staining and pitting of both type 304 and 316 stainless steels. Periodic cleaning of the steel panels, either manually or by natural rainfall, has been shown to minimise corrosion. It is concluded that both the type 304 and 316 compositions should provide acceptable corrosion resistance for offshore cladding applications, but type 316 is preferred because of its greater resistance to marine conditions, and the knowledge that conditions offshore are more severe than experienced in coastal installations.

In terms of mechanical properties, the austenitic stainless steels offer comparability with mild steel at ambient temperatures and considerable advantages at elevated temperatures, such that much thinner panel sections can be employed. These properties, combined with excellent fire resisting capabilities make the type 304 and type 316 compositions ideally suited to module cladding. Welding should present no difficulty as the techniques are well established, and in daily use in stainless steel fabricating shops around the world. When stainless steel is welded to carbon steel a slightly different technique is necessary but the methods are equally well proven.

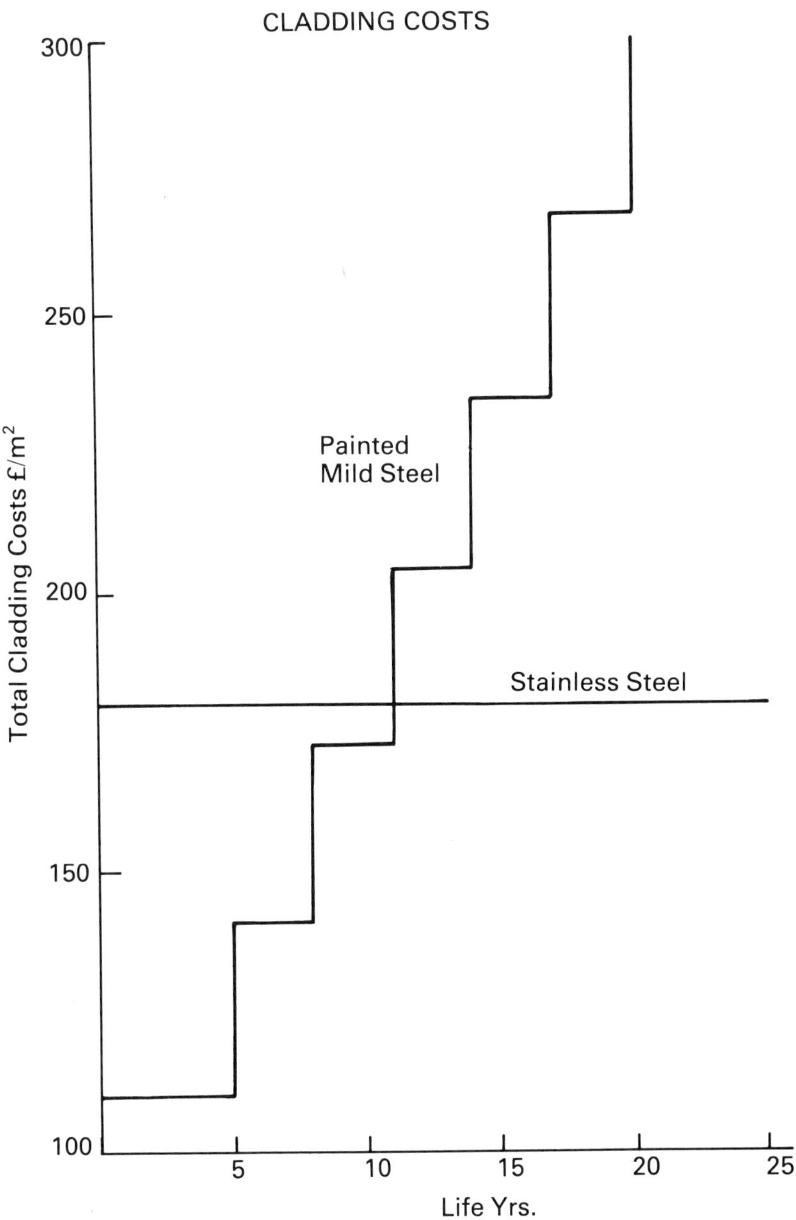

Fig. 2.6, 9 – Comparative costs of stainless steel and painted mild steel module walls.

8. COSTS

Stainless steel is, of course, more expensive than mild steel, or even mild steel that has been galvanised and painted. However, the more important costs are the additional costs of a completed module and the subsequent cost savings in maintenance.

In 1981, the cost of a typical painted mild steel panel was £110/m^2 whereas a stainless clad panel of similar fire rating would have cost £180/m^2. A typical accommodation module might have a clad area of 2000 m^2 and a total cost of £20 000 000. Consequently, the initial increase in capital costs was only ~0.7 per cent. On the credit side, the costs of maintenance were saved and the possibility of production interruption due to cladding failure or maintenance was removed. It is estimated that it costs about £500 per day for a maintenance painter and his equipment to be offshore. To scaffold, clean and repaint a module of 2000 m^2 could take 50 man-days, assuming good painting weather, which is not common in the North Sea. Assuming a cost of £15/m^2 for direct labour costs and consumables, e.g. blasting grit, paint, the maintenance cost is £500 $\times$ 50 man-days + £15 $\times$ 2000 m^2 = £55 000 or £63 333 if one day in four is lost due to bad weather.

The time to first maintenance will probably be 3–8 years, depending on the platform location and the quality of the original painting. Thereafter, the frequency could be less, say 3–4 years, because the quality of offshore workmanship will be inferior. Figure 2.6,9 shows that on the basis of these simple calculations, the stainless steel is better value after only eleven years of platform life. Of course, this has been a very simple approach but the benefits are so great that the advantage of stainless steel is still evident, however detailed are the economic calculations!

9. WEIGHT SAVINGS

Because there is no need for a 'corrosion allowance' on stainless steel clad modules, the steel thickness is much less than for mild steel, typically 2 mm compared with 4 mm. The actual weight of a typical A60 stainless steel panel is about 7 kg/m^2 less than mild steel, therefore, on a module of surface area of 2000 m^2, the weight saved is 14 tonnes. This can be used to improve platform efficiency by putting more equipment on the deck, or alternatively, a lighter support structure can be designed.

A saving of 14 tonnes topside saves a considerable weight of supporting structure, which is equivalent to a further capital cost saving of £140 000 or £70/m^2. This is the difference between mild steel and stainless steel initial costs and, therefore, it can be argued that stainless steel clad modules are cost effective from the moment that they are installed.

10. APPROVALS AND APPLICATIONS

Panels clad with stainless steel to BS 316S16 have been fire tested. Approval has generally been obtained, for example, the Defiant Cladding System has been approved by the certifying authorities and fire certificates have been awarded for the following ratings:

H120, A60H, A60, A0.

This means that the H120 panel will withstand hydrocarbon fire for over two hours.

Modules clad in stainless steel are currently in use on, or under construction for, the North Sea and other British coastal sites for the following platforms:

Field	*Operator*
Frigg	Elf/Total
Heimdol	Elf
Hutton	Conoco
Magnus	BP
Morecambe Bay	BGC
Murchison	Conoco
Nam	Total
North Alwyn	Total
Rough	BGC

11. OTHER OFFSHORE APPLICATIONS

Other applications of stainless steel offshore include: kitchen equipment, bunks, furniture, ceilings, toilets, etc. An offshore platform is, in effect, a floating hotel and many of the more luxurious stainless steel fittings that are found onshore are frequently used offshore. Stainless steel is commonly specified because it is important to ensure that the minimum number of components need replacing as the cost of transportation of equipment is very high.

12. CONCLUSIONS

Stainless steel clad modules are being increasingly preferred to painted carbon steel modules for offshore applications. This is because the North Sea marine environment is very corrosive and stainless steel can withstand many years of use without the need for costly offshore maintenance.

The stainless steel cladding system has been given full fire protection certification and the weight saving has led to substantial reductions in the size of the support structure.

The modular, pre-insulated, stainless system has now been proven and is being specified for an increasing number of offshore platforms.

13. ACKNOWLEDGEMENTS

The author would like to acknowledge the assistance of the International Nickel Company Limited for the information in section 4.2 and also BSC Stainless and Redpath Engineering Limited for general assistance.

The author wishes to thank Dr K. J. Irvine, Director, Research and Development, BSC, for permission to publish this paper.

2.7 CORROSION RESISTANT MATERIALS FOR SEA-WATER SYSTEMS

T. J. GLOVER, Phoenix Metals and Alloys Ltd., Redditch

ABSTRACT

The cost of maintenance of offshore structures to prevent the ravages of sea-water corrosion is rapidly becoming prohibitive. It is therefore economically sound to use materials which are corrosion resistant and require no maintenance, thereby allowing the platform to operate continuously, this is particularly applicable for critical components and systems. The properties and application of cupro-nickels, special stainless steels and high nickel alloys will be discussed with particular reference to piping systems and the splash zone.

INTRODUCTION

With the wide range of materials which are currently available virtually all corrosion problems can be technically resolved, it is, however, the economic considerations which are of more importance in this increasingly cost-conscious world we live in. Nowhere is this factor more vital than on offshore platforms where maintenance is extremely expensive – one oil company estimates that a job offshore costs 27 times more than it would onshore. In addition component failure can prove catastrophic both in cash and human resources. It is therefore essential to select materials for critical components which will resist corrosion and require little or no maintenance for the life of the platform. Cupro-nickel alloys and special stainless steels are considered to be ideal selections to withstand attack from marine environments.

STAINLESS STEELS

The traditional reaction of most engineers and designers to stainless steel is that it is not compatible with sea-water – stainless steel being used as a collective noun encompassing all grades as if they would all behave similarly. Stainless steel can be used in sea-water and its successful application depends on (a) service conditions and (b) the choice of alloy.

The most important environmental conditions are:

(i) *Velocity.* Conventional stainless steels should not be used in stagnant sea-water as there is strong possibility that pitting attack will occur but if a reasonable velocity can be maintained then these alloys can provide a reasonable service life. A good example of this is the use of Type 316 stainless steel impellers in sea-water pumps where the alloy exhibits outstanding resistance to impingement attack.

(ii) *Galvanic effects*. Stainless steel is generally more noble than the majority of metals and alloys used in marine engineering and will, if in direct electrical contact with a less noble material be galvanically protected. In sea-water pumps, for example, the casings are often made in Ni-resist cast iron and this affords protection to the stainless steel impeller and this is particularly useful during shutdown when the sea-water becomes stagnant.

(iii) *Aeration*. When sea-water is de-aerated it becomes much less aggressive and conventional stainless steels have been used with considerable success in desalination plants. Until recently carbon steel was widely used for the construction of flash chambers but corrosion rates in excess of 1 mm/yr have been observed [2.7,1]. The problem has been resolved by utilising solid and clad plate in Type 316 stainless steel for flash chamber wall construction. The sea-water is de-aerated to less than 50 p.p.b. and at this low level there is insufficient oxygen available to provide sufficient driving force for operation of the differential aeration cell which causes pitting in stainless steels. The lack of oxygen also reduces the incidence of stress corrosion cracking even though the plants operate at temperatures well in excess of 60 °C, the normal limiting temperature for safe use of conventional stainless steel in chloride ion environments. Many hundreds of tons of Type 316 stainless steel are now installed in multistage flash distillation units in the Middle East and to date have not presented any corrosion problems. The use of resistant alloys has enabled the plants to operate continuously, a far cry from the days of painted mild steel when the downtime for maintenance was often greater than the operating time.

(iv) *Marine environment*. Dr Hooper† has already described the savings in maintenance cost and weight which can be achieved by the use of stainless steel to replace galvanised mild steel. My own experience reinforces his opinion that Type 316 stainless steel will provide a surface which will remain intact and aesthetically attractive for the life of a platform. A small example of costs involving the consideration of alternative materials can be seen from Table 2.7,1. It can be clearly seen that the initial cost differential is small and that any maintenance cost over the platform life would immediately make the galvanised steel the less cost-effective choice. There is also the significant weight saving to be taken into consideration.

Table 2.7,1
Safety rail support

	Cost (£)	Wt. (kg)
Galvanised steel	13	8
316 stainless steel	20	3

†Previous paper, 2.6 in this volume.

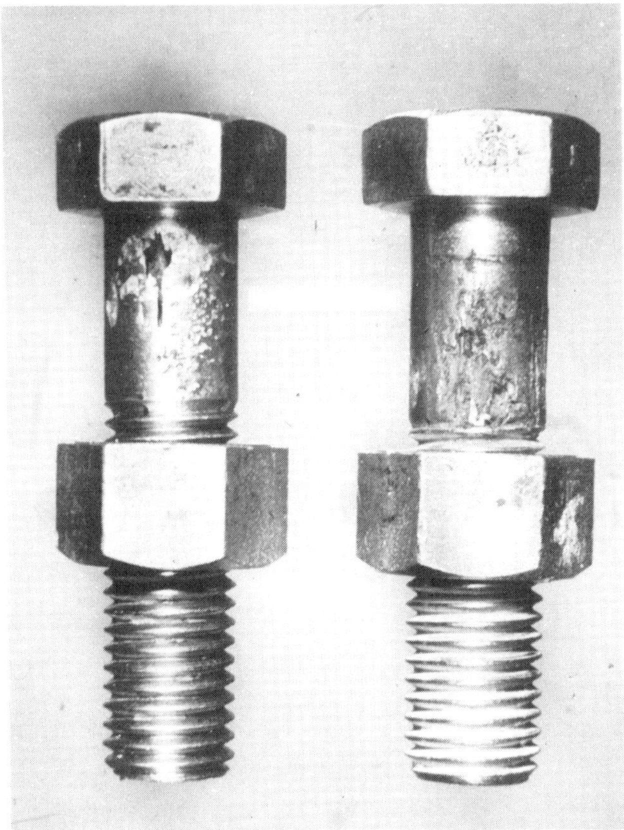

Fig. 2.7,1 – Type 316 both after 18 months' service in sea-water showing crevice attack under a nylon washer.

Under immersed conditions there is no doubt that conventional stainless steels are a problem and pitting and crevice corrosion are the most commonly encountered forms of attack particularly in stagnant conditions or under marine fouling. It is therefore evident that if the service conditions, i.e. velocity, aeration, etc. cannot be controlled then alternative grades of stainless steel must be considered.

AUSTENITIC ALLOYS

The alternatives available have been illustrated by Sedriks [2.7, 2] in his stainless steel family tree (Fig. 2.7, 2). It can be seen from this diagram that to improve the resistance to pitting corrosion the molybdenum and chromium contents must be increased. The addition of nitrogen is also useful as not only does it

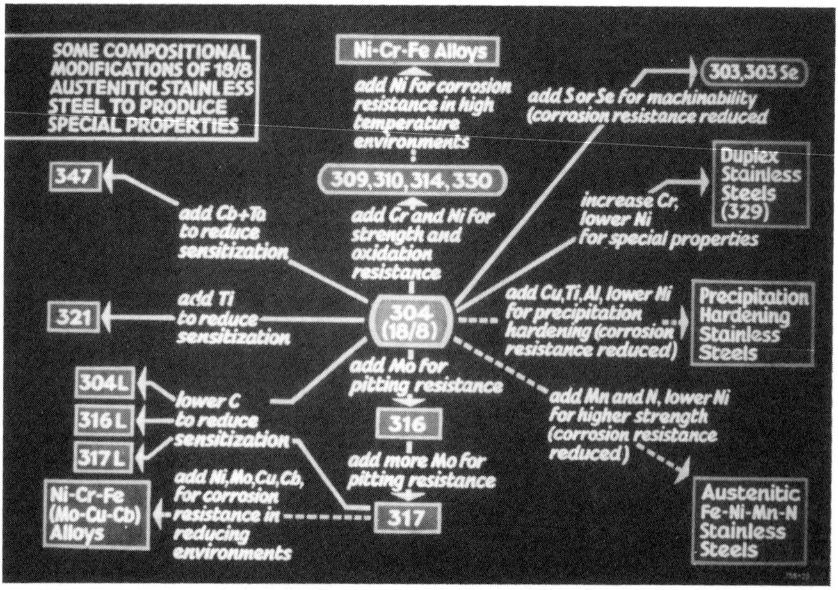

Fig. 2.7, 2 – Stainless steel family tree.

improve the pitting resistance it also increases the strength and stabilises the austenite.

A useful rule of thumb for assessing the chloride pitting resistance of an alloy is:

$$\text{pitting resistance} = \%\,\text{chromium} + 3.3 \times \%\,\text{molybdenum} + 16 \times \%\,\text{nitrogen}.$$

To obtain immunity from attack a value of 36 is normally recognised as being adequate. Values for some standard and proprietary grades of stainless steel are given in Table 2.7, 2.

Table 2.7, 2

Alloy	Pitting resistance value	
Type 304	18	Susceptible
Type 316	26	
Type 317	29	
904L	35	
SANICRO 28	38	Resistant
AL-6X	40	
254 SMO	42	

Laboratory and field experience would tend to confirm this rule of thumb and performance of the high chromium/molybdenum grades has been highly successful but no absolute guarantee of total immunity for an indefinite period can ever be made as there is always a possibility, albeit in this case extremely low, of corrosion occurring in some particular circumstance.

A level of 6 per cent molybdenum has been found to be the minimum level at which pitting and crevice corrosion resistance of stainless steel can be achieved using conventional chromium levels (~20%). Two such alloys have been independently developed in Europe and the USA. AL-6X was specifically developed for the sea-water condenser tubing in coastal power stations in the USA. It is normally available in thin wall longitudinally seam welded tubing only but because of its excellent resistance to pitting corrosion and impingement attack, 7 million metres have now been installed and are providing trouble-free service.

Fig. 2.7, 3 – AVESTA 254 SMO sea-water reject pipe (courtesy Maritime Service A/S).

Following successful trials in Scandinavia 64 kilometres of 254 SMO have been ordered for a sea-water cooled power station in Finland. 8–10 in. (203–254 mm) diameter pipes in the alloy have also been specified for the Golden Block field as an alternative to concrete lined mild steel for ballast piping to carry sea-water. Plate type heat exchangers and sea-water reject pipes (Fig. 2.7, 3) have already been employed on North Sea platforms.

FERRITIC ALLOYS

In addition to the alloy development on conventional austenitic alloys which has recently taken place there has also been considerable interest in developing ferritic grades which not only have the requisite resistance to pitting corrosion but which have good mechanical properties and which can be formed and welded. These alloys, usually referred to as the superferritics, were dicussed in considerable detail at a conference [2.7, 3] in Italy in 1980. Some of the alloys are already well established in sea-water condenser tubing. The most important grades are given in Table 2.7, 3.

Table 2.7, 3

Alloy	Composition %				
	Cr	Ni	Mo	Ti	C
AL 29-4C	29		4	0.4	0.02
Monit	25	4	4	0.4	<0.025
Sea Cure	26	2	3	0.5	0.02
Superferrit	28	3.5	2.5		<0.02

DUPLEX ALLOYS

Duplex ferritic/austenitic have been available for some time and offer higher strength, better stress corrosion, cracking resistance and in some instances superior pitting resistance compared to conventional austenitic alloys.

A low chromium grade typified by 3RE 60 (18 Cr, 4.5 Ni, 2.5 Mo) has been widely used in the petrochemical industry for heat exchanger tubing. A high chromium grade, Ferralium (25 Cr, 5.5 Ni, 3 Mo) has been used with considerable success for fasteners, shafts, pump and valve bodies in marine environments throughout the world.

An intermediate chromium grade (22 Cr, 5.5 Ni, 3 Mo, 0.15 N_2) developed in Germany and designated 1.4462 has now been added to the range. The alloy has been used for gathering lines and down-hole production tubing in some oil and gas wells in Holland and Germany. It has also been specified for production tubing in the Brae field. It can be seen from this brief review of some of the more important stainless steels which are commercially available from the vast range produced throughout the world, that many of the objections voiced against stainless steels can be overcome. Provided that the correct selection is made then many existing corrosion problems can be minimised and even eliminated, often at little additional expense.

CUPRO-NICKELS

Whereas special stainless steels are a recent innovation on offshore structures cupro-nickels are already well established and have been widely used for piping systems. Their selection is based largely on the low corrosion rates observed for the alloys in sea-water flowing at moderate velocity. An equilibrium corrosion rate of 1.3 μm/year was established in sea-water flowing at 0.6 m/sec over a 14-year period [2.7, 4] (Fig. 2.7, 4).

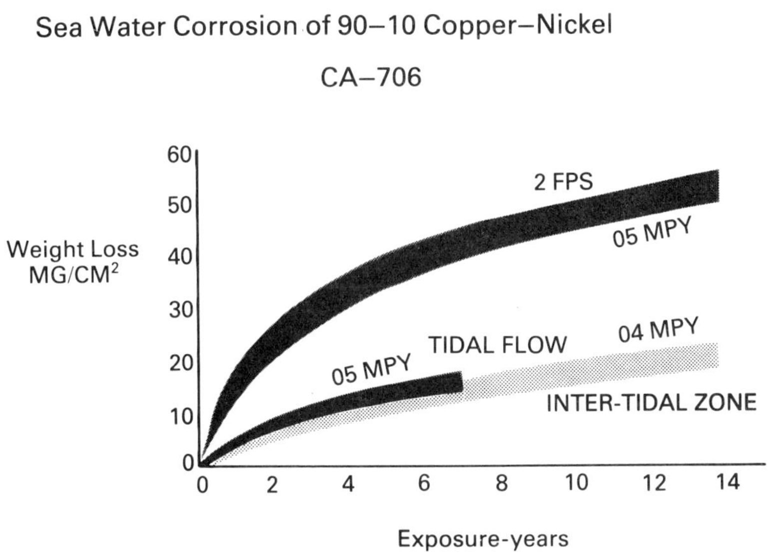

Fig. 2.7, 4 – Fourteen-year data for cupro-nickel.

A range of cupro-nickel alloys is available but by far the most commonly encountered is the 90/10 grade (ASTM alloy No. C70600) and unless otherwise stated all discussion in the paper will refer to this particular grade.

Initially sea-water piping systems were constructed from bare, cement lined or epoxy lined carbon steel. It was soon evident that the protective coatings were prone to break-down and serious corrosion problems occurred within 2-5 years [2.7, 5]. Conversion of many of these systems to cupro-nickel was found to be not only a technical solution but also extremely cost effective due to the lack of maintenance required. Today curpo-nickel is specified on a regular basis for sea-water pipe-work and is operating successfully on Claymore and Piper (Occidental), Ninian (Chevron), Auk, Cormorant and Fulmer (Shell), Tartan (Texaco), Frigg TCP 2 (Elf) and Murchison (Conoco).

Seamless pipes and fitting are readily available from UK suppliers up to 400 mm diameter (Fig. 2.7, 5). Much larger diameters, however, can be obtained using seam welded construction and since the installation of a 1350 ton brake press in Gloucester diameters up to 1200 mm and lengths up to 8 m can be obtained.

Fig. 2.7, 5 – Seamless pipe and fittings for Ninian field.

As a general rule the use of cupro-nickel has proved highly effective except where the limiting design velocity has been exceeded. BS MA18 specifies that this velocity is 3.5 m/sec for pipes 100 mm in diameter or larger, above this velocity there is a probability of impingement attack which normally manifests itself at bends or downstream of valves. The probability of impingement attack by sea-water has been well illustrated by Gilbert [2.7, 6] (Fig. 2.7, 6). A marginal improvement can be obtained by using the 70/30 grade of cupro-nickel but if a significantly greater velocity capability is required in an existing system then Monel 400 would be ideal as it can tolerate 40 m/sec and is galvanically compatible.

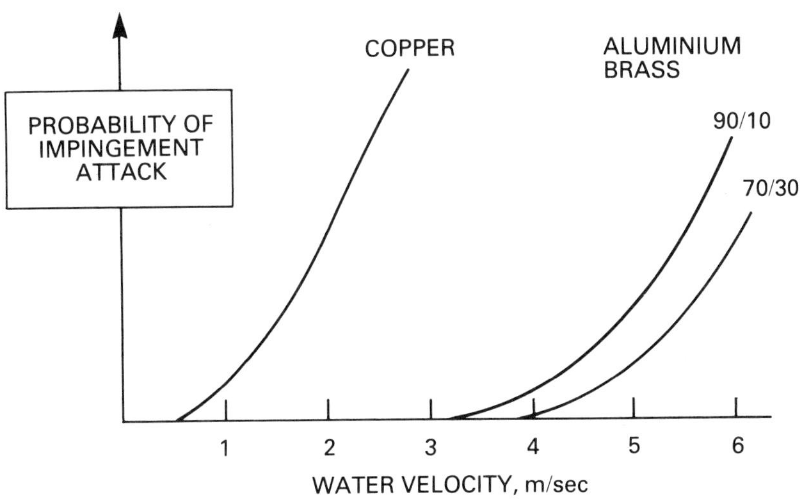

Fig. 2.7, 6 – Velocity effects on copper alloys.

A more minor problem is galvanic corrosion between cupro-nickel and less noble metals when placed in direct electrical contact in a continuous electrolyte and is most commonly encountered when cupro-nickel pipes are connected to ferrous valve or pump bodies. The problem has been overcome or minimised by the use of (a) paint coating on the cupro-nickel cathode, (b) a sacrificial spool piece of Ni-resist cast iron interposed between the disimilar metal couple or (c) insulating non-metal jackets and sleeves (in practice a complicated, not completely reliable procedure).

In addition to the low corrosion rates observed on cupro-nickel alloys in moderate velocity sea-water they have the added advantage of being naturally anti-fouling. Copper alloys having more than 70 per cent copper deter settlement of marine plant and animal organisms by the biocidal action of copper ions.

Initially this is due to the release of copper ions into the surrounding water but gradually a surface oxide film is formed which hydrolyses to produce an outer layer of copper hydroxychloride. This film is only loosely adherent and will become readily detached by the weight of any biofouling which settles on it, thus exposing a fresh biocidally active surface. In this way copper nickel can maintain its anti-fouling resistance indefinitely provided that it is allowed to corrode, albeit very slowly, in a natural manner. The anti-fouling properties can be seen by a comparison of panels of steel, copper and cupro-nickel which have been immersed in Portsmouth harbour for 4 years (Fig. 2.7,7).

Fig. 2.7,7 – Four-year exposure in Langstone Harbour, Portsmouth.
L–R (steel–copper–90/10 cupro-nickel).

This anti-fouling characteristic has been used to good advantage in other marine applications the most notable being:

Ships' hull construction

Over the last decade several cupro-nickel hulled vessels have been built, the most well known is the *Copper Mariner* [2.7,7] launched in 1971. Throughout the past 11 years the hull has remained free from fouling and has never required any maintenance. The condition of the hull after 4 years compared with that of its sister ship after only 6 months which had a conventional painted steel hull, can be seen in Fig. 2.7, 8. The corrosion rate on the hull was only 0.05 mm after 4 years indicating a virtually indefinite life for the 6 mm thick hull plate.

Fig. 2.7, 8 – *Copper Mariner* and *Asperida*.

Fish farming

Conventional fish cages are constructed from galvanised iron wire or plastic and rapidly become fouled which increases the weight and drag of the cage, reduces water flow, and thereby oxygen to the fish, and requires regular cleaning maintenance. Experience in the USA with a prototype INCRA cage [2.7, 8] has shown that by using cupro-nickel expanded metal the cage remained free from fouling and the fish stock yield was significantly improved. In addition the cage is more durable being resistant to storms and predators such as seals.

The success of the INCRA cage trials has led to the commercial use of cupro-nickel in Scotland, Scandinavia and North America and the conditon of a cage being operated off Bergen in Norway can be seen in Fig. 2.7, 9. The cupro-nickel is free from fouling whereas the fibreglass frames are heavily fouled.

Cupro-nickel mesh offers similar advantages in shellfish farming and currently 40 cages are on trial in Spain where they are suspended from a raft so that they can be easily raised during harvest time.

Sea-water intakes

Sea-water intakes to coastal power stations are normally constructed from steel or concrete and are susceptible to heavy fouling. This reduces the effective diameter of the pipe and increases pumping costs. A further problem is that when the fouling, particularly the hard shelled mussels and barnacles, become

Fig. 2.7, 9 – Bergen fish cage.

Fig. 2.7, 10 – 3000 mm diameter cupro-nickel sea-water intake.

dislodged they can find their way in to the heat exchanger systems, causing blockage and initiating impingement attack. Chlorination has until recently been the normal technique for minimising the amount of fouling on the intake walls. An alternative technique is to use cupro-nickel intakes which require no chlorination treatment. The first [2.7,9] of these intakes was installed in the Middle East in 1981. The submerged section was constructed from 6 mm solid 90/10 cupro-nickel clad plate (Fig. 2.7,10) and the overland section from 15 + 2 mm 90/10 cupro-nickel roll clad plate.

Cladding offshore structures

As biofouling is a major problem [2.7,10] on offshore structures consideration has been given to determine if the unique properties of cupro-nickel can be utilised to reduce the need for regular cleaning of the submerged structures. Not only would this save money by reduced maintenance but it would also ensure that the original design for loading and clear passage of sea-water through the steel lattice work of the platform would be maintained.

Metal cladding [2.7,11] in the splash zone with Monel 400 is a well established technique in the North Sea; attachment is made by welding [2.7,12]. Cupro-nickel is now being specified for the same application. Experience suggests that it will provide the necessary corrosion resistance to give a long service life. If, however, it is attached to the steel structure by welding in a similar manner to that used for Monel 400 its anti-fouling properties will be seriously impaired, if not completely destroyed. This is because the weld would result in a galvanic cell being set up in submerged conditions in which the steel would corrode sacrificially thereby protecting the cupro-nickel and preventing it corroding naturally. Similarly, the application of conventional impressed current or sacrificial anode cathodic protection would also prevent natural corrosion of the cupro-nickel and allow fouling to occur.

It is therefore essential that some form of insulation is applied between the steel structure and the cupro-nickel cladding if the unique properties of cupro-nickel are to be utilised. The use of cement grouting and heavy neoprene rubber linings is currently being evaluated.

REFERENCES

[2.7,1] Oldfield, J. W. and Todd, B., 'Corrosion considerations in selecting metals for flash chambers', *IDEA Int. Conf. on Desalination and water reuse*, Nice, Oct. 1979.

[2.7,2] Sedriks, A. J., 'Corrosion of stainless steels', *Corrosion Monograph*, Wiley, 1979, New York.

[2.7,3] 'Advanced stainless steels for sea water applications', Climax Molybdenum/Associazione Italiana di Metallurgia Symposium, Piaceza, Italy, Feb. 1980.

[2.7,4] Efird, K. O. and Anderson, D. B. 'Sea water corrosion of 90/10 and 70/30 cupro-nickel, 14 year exposures', *Materials Performance*, 1975, **37**, 37-40.

[2.7,5] Gilbert, P.T., *3rd Int. Conf. on Present and Future Markets for Copper*, London, Oct. 1981.

[2.7,6] Gilbert, P. T., *Proc. Inst. Marine Engrs material Selection Symposium*, March, 1968, 14.

[2.7,7] Tuthill, A. H. *et al.*, 'CA 706 copper nickel alloy hulls – the Copper Mariner's experience and economics', *Soc. of Naval Architects and Marine Engineers Conf.*, New York, Nov. 1976.

[2.7,8] Huguersin, J. E. *et al.*, 'Experiences with a fouling resistant modular marine fish cage system', *Proc. Bio-Engineering Symposium for Fish Culture*, Oct. 1979.

[2.7,9] Jasner, M., 'Copper in marine engineering – oil rigs and platforms: salt water lines', *3rd Int. Conf. on Present and Future Markets for Copper*, London, Oct. 1981.

[2.7,10] Freeman, J. H. Marine fouling of fixed offshore installations, *DOE Report OPTI*, Sept. 1977.

[2.7,11] Ross, R. W. and Anderson, D. B. 'Protection of steel piling in marine splash and spray zones – the metallic sheathing concept', *Proc. 4th Int. Congress on Marine Corrosion and Fouling*, Juan-les-Pins, Antibes, June 1976.

[2.7,12] McKeown, D. 'Monel alloy 400 sheathing – protection of offshore structures in the splash zone', *Metal Construction*, Dec. 1976.

2.8 APPLICATIONS OF NOVEL ANTI-FOULING COATINGS ON OFFSHORE STRUCTURES

D. MILLER, A. M. CAMERON and **E. B. SHONE**, Shell Research Limited,
Thornton Research Centre, Chester

ABSTRACT

A research programme, undertaken to assess the effectiveness of conventional anti-fouling paints, is described. The results have indicated a need for alternative approaches to the problem of long-term protection from fouling for offshore structures. Cu/Ni cladding has long been advocated as a possible solution and this is considered. A non-toxic coating being developed by Shell Research Ltd is described and this may represent a significant advance in the field.

INTRODUCTION

Offshore structures become fouled by marine growths, the extent and type being largely dependent on geographic location. In general, the area of the platform that is subjected to the most severe fouling is situated between sea-level and a depth of approximately 30 meters. Fouling significantly increases the structure's weight, its surface roughness and the diameter of its legs and members. This in turn increases the hydrodynamic loading on the part of the structure that is subjected to the highest wave loadings and hence reduces its stability. Additionally, various forms of corrosion (crevice and bacterial) can occur beneath the fouling. Visual inspection can become difficult or impossible. Thus, when the degree of fouling is greater than anticipated, design specifications can be rapidly over-extended and regular inspection and removal of the fouling must be carried out.

Design specifications could be modified such that more fouling may be tolerated before the stability of the structure is threatened. This would be costly and would not solve all of the problems associated with crevice and bacterial corrosion which can occur beneath the fouling. Additionally, inspection procedures would become increasingly difficult to conduct. The ideal solution to this problem would be to reduce, significantly, the extent of fouling that is occurring.

CONVENTIONAL ANTI-FOULING PAINTS

For many years we have evaluated conventional anti-fouling coatings used by, or available to, fleet operators. These paints aim to prevent the settlement of fouling spores and organisms by the continuous release of toxic substances into the water layers adjacent to the coated substrate. The mechanism by which these toxins are released was uncertain but was thought to be strongly dependent on the composition of the toxin and binder components of each paint formulation.

Ideally, careful selection of these components would result in liberation of the toxin at a controlled rate and so allow predictions of the life expectancy of the paint to be made.

In practice, it was found that many anti-fouling paints failed prematurely in service. Frequently they were found to contain more poison than should have been necessary to afford protection. Obviously the initial leaching rate was not being maintained but the reasons for this were not clear. Laboratory exposures

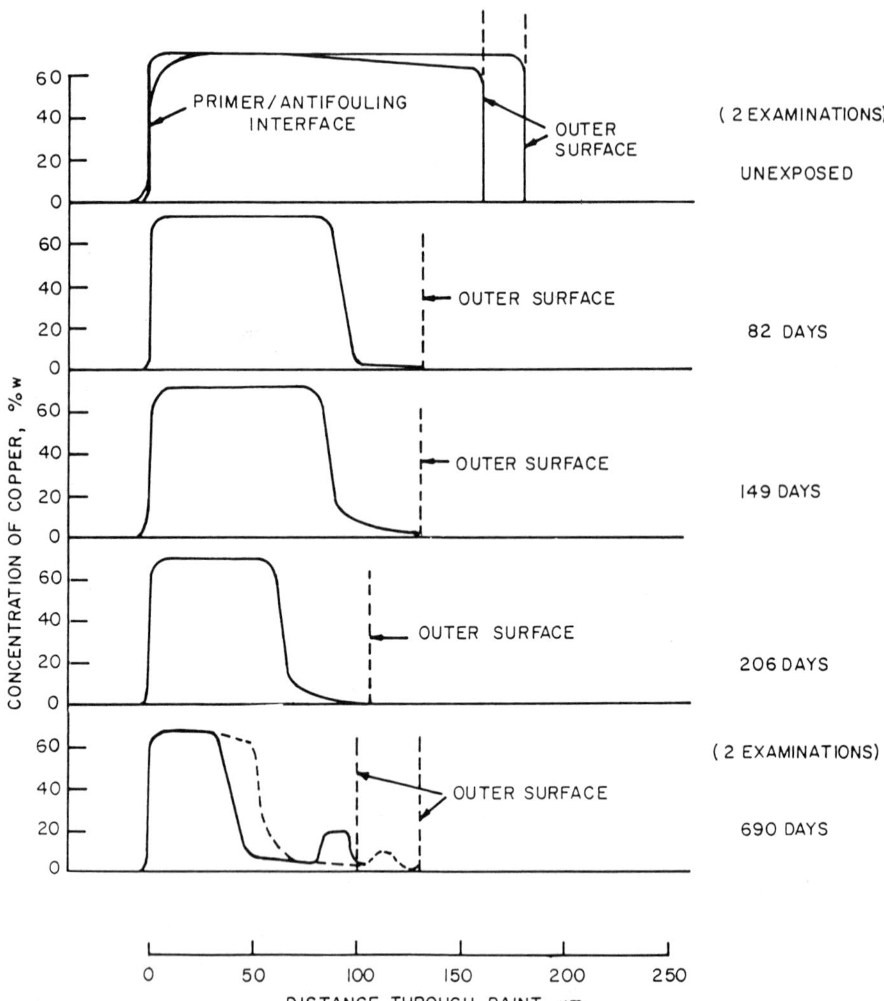

Fig. 2.8,1 – Large test patches applied to the ship's hull for regular visual inspection.

and static raft trials conducted by numerous authorities were unable to clarify the situation. Consequently an evaluation programme was carried out on panels exposed to more realistic conditions in an attempt to explain this apparent anomaly.

A series of trials have been conducted in which large patches of anti-fouling paints were applied to ships' hulls. To complement these patches and to facilitate regular inspection of the paints several small panels of every formulation of interest were also attached to the hulls (Figs 2.8,1 and 2.8,2). These panels could easily be removed, as required, and returned to the laboratory for evaluation. The techniques of sample preparation and anlaysis adopted for this purpose were not simple. Great care had to be taken both when preparing the samples for examination and when interpreting the results obtained by scanning electron microscopy. The techniques, which were to prove most successful, have been described elsewhere ([2.8,1], [2.8,2]).

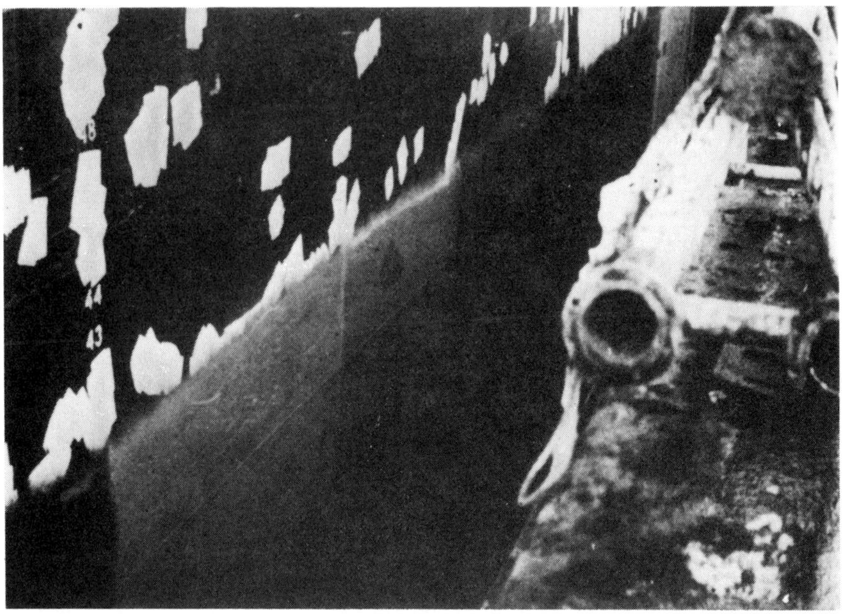

Fig. 2.8, 2 – Small removable panels attached to the ship's hull for laboratory investigation.

A typical result obtained from this investigation is presented in Fig. 2.8,3. This shows the variation of copper concentration through a paint film as a function of time. It can be seen that a surface layer that is depleted of copper is rapidly formed. This increases the distance through which copper ions must be transported, and difficulty is experienced in maintaining a toxic concentration

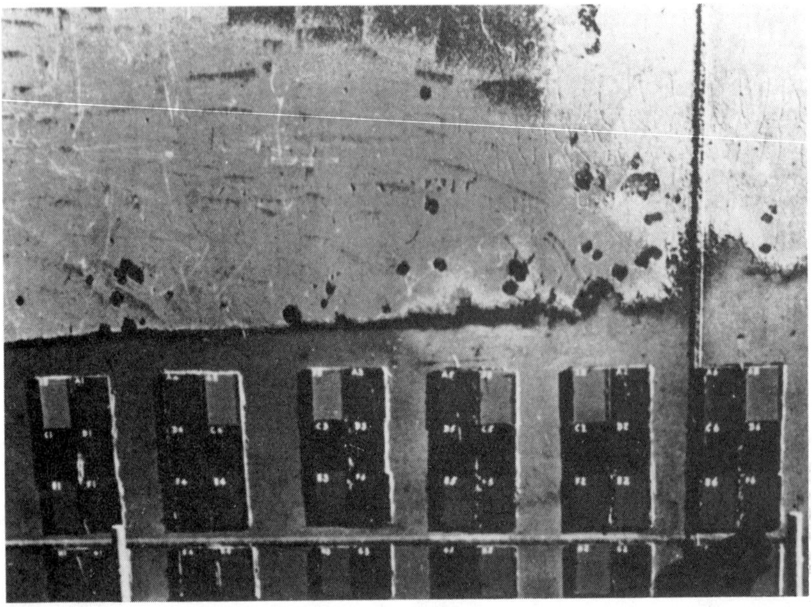

Fig. 2.8, 3 – Variation of toxin distribution and concentration in a typical copper-based anti-fouling coat.

at the paint/water interface. Premature fouling can then take place. Another interesting feature that can be seen from Fig. 2.8, 3 is the formation of a small peak in copper concentration at the paint/water interface after 690 days. This has been shown to be associated with insoluble copper hydroxychlorides which will, depending on the extent of their formation, partially or totally prevent any limited leaching of toxin that may still be occurring. Experience has shown that under these circumstances fouling can be severe.

It should be noted that similar results were obtained for organotin-based anti-foulants. Denuded zones were again formed in a relatively short period and whilst there was no detectable formation of insoluble salts, the rapid leaching rates observed mean that this type of toxin offers no real advantage over copper oxide with regard to life expectancy. In general, systems of this type will only afford protection for up to twelve months.

SELF-POLISHING COPOLYMERS

In recent years, major advances have been made in the development of a family of paints known as self-polishing copolymers (SPC). These paints have significantly extended the duration of active anti-fouling performance. This is largely due to the incorporation of the toxins in a matrix that is sparingly soluble in sea-water. Consequently there is no build-up of denuded layers, the matrix continuously dissolving to expose a fresh paint/water interface. The application of SPC on to offshore platforms is limited by two considerations. First, the best of this new variety of paints is still only capable of protection for a period of time that is considerably shorter than the service life of an offshore platform. Secondly, these systems were devised for use on ships where erosion of the matrix will be enhanced by a relatively rapid flow of water over the surface. It is unlikely that wave and tidal action alone would be sufficient to ensure the degree of dissolution required.

COPPER/NICKEL CLADDING FOR OFFSHORE APPLICATION

Copper sheathing has been used as protection against fouling on wooden ships for several hundred years. The protection mechanism is simple, and depends upon the slow dissolution of copper, producing Cu ions, which in turn prevent fouling. Several modern small ships have been constructed from a 90/10 copper/nickel alloy. These small ships are claimed to have performed well in service maintaining a hull free of fouling with subsequent reduction in operating costs [2.8, 3]. The 90/10 Cu/Ni alloy from which they are fabricated erodes significantly when water speeds of 4 m/s or greater flow across its surface [2.8, 4]. Thus there is no build-up of insoluble copper salts.

Many modern ships' heat exchangers are manufactured from 90/10 Cu/Ni alloy and are designed to operate with water velocities of less than 2 m/s, thus avoiding serious erosion/corrosion damage. These heat exchangers have to be protected from fouling and various techniques are used, including the electrolytic generation of chlorine. During examinations of units, fouling has been observed and crevice corrosion under barnacles attached to the tube material has been identified as the source of failure.

Offshore structures often stand in relatively stagnant waters with tidal flows of less than 2 m/s. It is possible therefore that the formation of insoluble salts of copper on the surface of the alloy could inhibit the formation of copper ions, thus allowing fouling to occur. A 90/10 Cu/Ni panel has been exposed in the open sea for the past two years off the south coast of England (Fig. 2.8, 4). At present it is free of fouling but covered with a green deposit which is most probably $Cu_2(OH)_3Cl$. Future inspections will determine the effectiveness of the material as a long-life anti-fouling coating. Another possible limitation is that Cu/Ni cladding material is noble with respect to the steel jackets to which it would be applied. Thus great care must be taken to ensure that no galvanic couples are created during its fabrication, as this could lead to increased corrosion of the steel. Additionally, direct electrical contact between the Cu/Ni and steel, subsea, where cathodic protection is operative, could lead to protection of the cladding and subsequent loss of its anti-fouling performance.

Fig. 2.8, 4 – A 90/10 copper/nickel panel exposed on a test raft for two years had no significant fouling attached.

NOVEL ANTI-FOULING COATINGS FOR OFFSHORE APPLICATION

An alternative anti-fouling system is being developed that is nontoxic, depending upon physical rather than chemical properties for its performance. The system is manufactured from silicone-rubber-based materials to which an exuding fluid is added. All of the materials are hydrophobic, which reduces the ability of the marine organisms to form bonds with the coating's surface. Additionally, the fluid slowly exudes to the surface, constantly renewing it, and enhances the anti-fouling properties.

Several materials, applied 200 μm thick on to test panels, have been exposed on a raft facility for the past four years. They have remained free of fouling throughout the exposure period (Fig. 2.8, 5).

Dexter [2.8, 5], and Fletcher and Loeb [2.8, 6], have identified relationships between the critical surface tension of a substrate and the rate of attachment of marine bacteria to it. Practical trials with low critical surface tension coatings carried out by Griffiths and Bultman have indicated that Teflon has some anti-fouling properties but they concluded that 'the most effective materials were the polydimethylsiloxanes' [2.8, 7].

Fig. 2.8, 5 – Two silicone based materials, 200 μm thick, that have been exposed on a test raft for four years.

Following successful raft trials and support for the general principle in scientific publications, several coatings have been adapted for offshore applications. As the performance of the coatings is improved by additions of hydrophobic fluids, which exude to the free surface, it is believed that their life expectancy is a function of the reservoir of fluid within the coating. By increasing the coating thickness the quantity of fluid available per unit area of exposed surface can be increased and along with it the effective life of the coating. Additionally new materials, containing a reinforcing network of organic polymer that markedly improves the mechanical properties of pure silicone rubber, have been developed. A pipe section coated with two pure and three reinforced silicones has been exposed at a site in the Pacific Ocean. Two inspections have confirmed the anti-fouling performance of the coating (Fig. 2.8, 6). The unprotected member on to which the pipe section was attached had doubled in diameter within 10 months while the novel coatings were virtually free of fouling after two years. A similar exercise has been carried out on platforms in the southern North Sea and in the China Sea (Fig. 2.8,7). In both of these locations novel anti-fouling materials are peforming well.

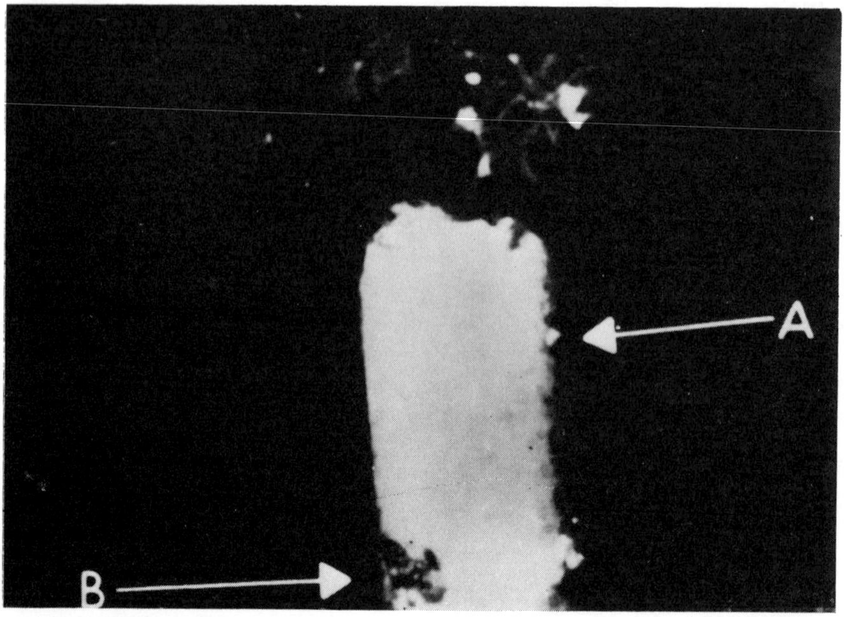

Fig. 2.8,6 – A view of two anti-fouling coatings after 10 months' exposure in the Pacific Ocean. Note that mussel growth has attained a thickness of 12 cm on the unprotected structure.

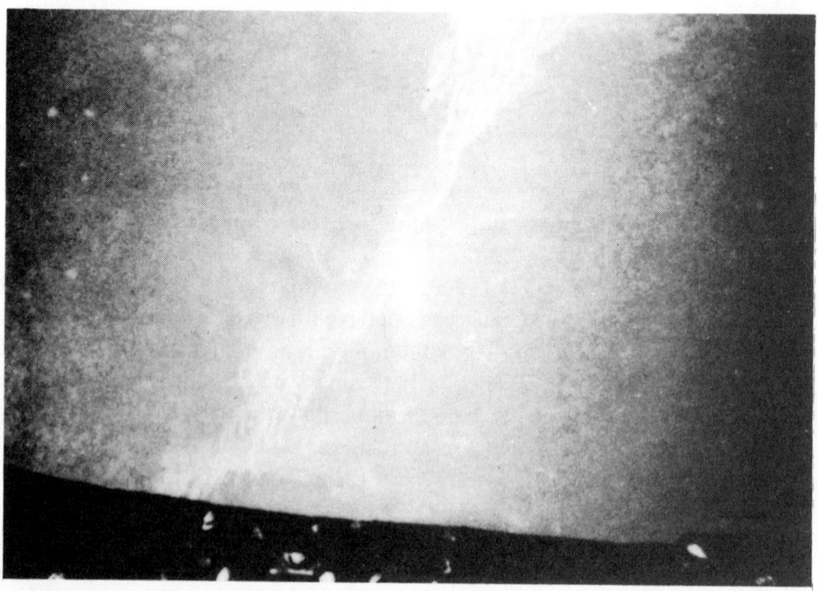

Fig. 2.8,7 – 46.5 in diameter tubular on to which a novel anti-fouling coating was cast. The lower uncoated area is fouled with barnacles after 3 months in the China Sea.

THE DEVELOPMENT OF A SUBSEA LOCATION MARKER

The need for a system of subsea location markers was recognised at Thornton. Large featureless concrete platforms and steel jackets on which many of the nodal structures are repeated, pose orientation problems for divers wishing to relocate areas of interest. Conventional paint marker systems are quickly obscured by marine growths. By utilising the novel anti-fouling material and including some modifications an anti-fouling subsea location marker has been developed. High contrast characters are readily produced by incorporating various pigments into some areas of the markers. Several have been manufactured containing fluorescent pigments that are sensitive to ultraviolet light, and may offer advantages when unmanned submersibles are used for inspection. Several methods for the attachment of markers are proposed, including a simple buckle and belt arrangement or, by first securing them in glass-reinforced plastic, they can then be bolted on to the structures. The versatility of the marker material allows complex characters to be produced (Fig. 2.8, 8).

Fig. 2.8, 8 – Subsea location marker, prior to installation on an offshore structure.

CONCLUSIONS

1. Conventional anti-fouling coatings have a life expectancy of approximately one year. They do not, therefore, offer a solution to the problems associated with fouling on rigs, which may be in service for 20 years or more.
2. Cu/Ni cladding may be a successful long life anti-fouling material; however, it is expensive. Additionally, problems associated with bi-metallic couple formation and observed inconsistencies in the anti-fouling performance require further elucidation.
3. Coatings derived from silicone materials have remained free of fouling for periods in excess of four years and as such may, in thick sections, represent a credible solution to the problem of fouling on offshore structures.
4. Subsea location markers manufactured from silicone materials could offer a versatile and economical means of finding specific locations of the submerged areas of offshore structures.

REFERENCES

[2.8, 1] Bird, R. J., *J. O. C. C. A.*, 1977, **60**, 256-262.

[2.8, 2] Bird, R. J. and Park, D., *J. O. C. C. A.*, 1978, **61**, 151-156.

[2.8, 3] Moreton, B. B. and Glover, T. J., *5th International Congress on Marine Corrosion and Fouling*, 1980.

[2.8, 4] Tuthill, A. H. and Schillmolle, *The Ocean Science and Engineering Conference*, Washington DC. 1965.

[2.8, 5] Dexter, S. C., *Proceedings 4th International Congress of Marine Corrosion and Fouling*, 1976, 131-138.

[2.8, 6] Fletcher, M. and Loeb, G. I., *Colloid and Interface Science*, 1976, **V**, 459-469.

[2.8, 7] Griffiths, J. R. and Bultman, J. D., *J. Naval Engrs*, 1980, **92**, 129-132.

CHAPTER 3

Discussion

A wide ranging discussion took place on various points raised on the presented papers and on related matters. Some of these are summarised in the following account of the discussion,

On Paper 1.2 (Patterns of macrofouling on steel platforms in the central and northern North Sea) Mr A. N. Walker (Robert Gordon's Institute of Technology) asked if mussels and barnacles appeared on cathodic protection anodes of (a) the sacrificial, or (b) impressed current type, or (c) both types. Dr Forteath said that they appeared on (a). Dr T. G. Gooch (Welding Institute) asked whether ferrous sulphide (FeS) had been found under corrosion products beneath fouling and, if so, how was it identified. Dr Forteath said that FeS in such situations had been detected using a scanning electron microscope with EDAP facility in which the trace had been compared with that of a known standard of FeS.

On Paper 1.3 (Sulphate-reducing bacteria and anaerobic corrosion) Dr R. Worthington asked Professor Hamilton if there was any evidence that acidic bacterial metabolites are a source of corrosion problems. Professor Hamilton replied that there is some suggestion that acidic microbial growth can play a part in some corrosion processes which are in any case extremely complex. Under the special conditions of anaerobosis, however, associated with sulphate-reducing bacteria, it seems certain that the only two significant mechanisms are the removal of hydrogen from the cathode and/or the formation of sulphide films. Mr E. C. Faulds (Shell Expro, Aberdeen) asked what was the anode material used to give cathodic protection to the coupons used. Professor Hamilton replied that the anode material on Brent and Brae was US military specification 18001 producing 780 amp.h/kg. In answer to Dr Hardy's (B.P., Sunbury) question on the potential of the protected steel he explained that the anodes were designed to give 135 mA/m^3 which was to Det Norsk Veritas standards for northern North Sea activity which should result in ~ -850 mV potential with respect to the Ag/AgCl reference electrode.

In the discussion on Paper 1.5 (Influences of microalgae on corrosion of structural steel) Dr J. Lewis (University of Aberdeen) asked for the sequence of events in the development of marine fouling on a structure. Dr Terry provided the following reply. 'The initial fouling sequence of any substrata immersed in the sea can be divided into three phases [3.1,1]. Attachment of common chemo-ogranotrophic bacteria, which readily utilise the organic nutrients adsorbed on to the substratum is followed by the attachment and growth of more nutritionally specialised stalked and filamentous forms. Thirdly, diatoms, other microalgae and protozoans settle in large numbers to be eventually followed by microalgal and invertebrate fouling. We have observed bacteria to settle in large numbers within a few hours of immersion of test plates while diatoms can reach large numbers in from 3 to 7 days and a considerable "slime film" can be built up within a month. While the sequence in settlement usually follows this pattern the time scale and species involved will depend on the season and geographical position of the substratum'. In another contribution to the discussion on this paper Mr J. A. Bray (MATSU, Harwell) pointed out that the authors had described how the initial microalgal layer on a cathodically protected surface would be virtually encapsulated in calcium carbonate. In view of this it seemed unlikely that the pH levels quoted of about 1.5 could be achieved or maintained. In reply to this point the authors emphasised that acid-producing bacteria can be very localised and active and thus in a small encapsulated microalgal inclusion in a scale the decomposition process could well produce very low pH values. The acid conditions will act on the calcium carbonate which will begin to go into solution at about pH 6.5. At sufficiently low pH values the carbonate will dissociate into carbonic acid or carbonic acid and water. Obviously the effect will depend on the amount of organic material and the conditions at the site and complications will arise once the scale had been removed and the cathodic products produced at the steel surface by the cathodic protection systems interact with the acid production. On substrata such as unprotected steel or paint films the effects of low pH are more easily envisaged.

There was some concern expressed on the use of a zinc electrode for measuring potentials of cathodically protected specimens that was reported in Paper 1.6 (Fouling and corrosion off the Outer Hebrides). Mr A. Mercer (NPL, Teddington) asked whether this gave reliable results. The authors thought that the method was reliable but Dr T. G. Gooch (Welding Res. Inst.) and Dr P. E. Francis (NPL) expressed further reservations.

Mr B. Balmer, the author of Paper 1.1 (Underwater inspection for corrosion) was asked two questions by Mr D. H. Swallow of B. P. Pet. Dev. Ltd, Dyce, (i) 'Was access available to all parts of the structure and particularly the deep central part (in view of the need to be selective on all parts of the platform to be inspected)?' and (ii) 'Were there any problems in operating a saturation diving spread supported from the platform?'. The reply to the first of these was 'Generally no; the Ninian Central platform inside of the Jarlan wall No. 2 has required the installation of a platform-based diving system for this purpose', and to (ii) 'No; in fact there are many positive advantages in operating from the platform'. Mr T. J. Glover (Pheonix Metals, Redditch, Worcs) enquired whether Chevron – as the only operator with Monel 400 sheathing in the splash zone on legs, as opposed to risers – had experienced any preferential attack at the Monel/steel interface. 'Had the Monel or steel surfaces been painted?' The reply was that no such attack had been noted and that these areas are inspected annually. The corrosion protection is afforded by coating and cathodic protection. The area of the Monel/steel interface is coated for approximately one metre top and bottom of the interface weld.

On Paper 2.3 (Mixed metal galvanic problems associated with offshore sea-water pumping equipment) Mr Swallow asked Mr Robson to comment further on the technique that he had described of using carbon steel backing flanges and epoxy coatings for Cunifer sea-water pipework and whether any consequential effects had been observed when coatings had not been used. The author, in reply, indicated that the data reported in his paper were based on experience on a platform which has a number of such flanges and which has been operating in the North Sea for about two years. Experience had shown that the epoxy coating was prone to damage and that, as a consequence, the electrical insulation between the cupro-nickel pipework and the carbon steel backing flanges had rapidly deteriorated. However, the backing flanges used on this platform had been galvanised prior to coating and therefore very little corrosion of the backing flange was evident. If the flanges were not galvanised galvanically-accelerated corrosion would undoubtedly occur and it is possible that over a period of years the build-up of corrosion products between the backing flange and the pipe stub flange would be sufficient to strain the flange bolts beyond their limit. This problem could be overcome by using high strength, grade 8.8 bolts in conjunction with backing flanges of grade 43A steel. In such an arrangement the backing flange would deform before bolt failure could occur. Mr Robson went on to say that a more suitable coating, which is less prone to impact damage, would have been one of the elastomeric polyurethanes, and furthermore that this could have been applied to only one side of galvanised backing flanges. In effect, such a coating would have acted as a gasket glued to the mating face of the backing flange. However, in view of the cost of coating large numbers of backing flanges and the extensive quality control required to ensure a high standard of coating, it would probably be cheaper and easier to insert a split-ring rubber gasket between the galvanised backing flange and the stub flange of the cupro-nickel pipe.

Dr Lewis asked whether in view of the considerable corrosion occurring in mixed metal pumping systems there was a new generation of pumping equipment being developed. Mr Robson was able to confirm that the sea-water pumping equipment presently being offered is infinitely superior to that which was on offer four or five years ago. Improvements both in materials slection and in the combination of materials are very much in evidence in the new generation of pumping equipment. Nevertheless, because of cash flow restrictions, the old designs, using cheaper, less durable materials are often used on new projects, even though the estimated lifetimes of such equipment are generally agreed to be only four or five years. In mitigation of this policy the combination of materials in these old designs has generally been modified to avoid many of the glaring galvanic problems experienced in previous models. Dr R. Hooper (BSC, Rotherham) referred to the author's comments on coating a two-foot length of 12 in diameter stainless pipe connected to mild steel pipe to prevent galvanic corrosion. 'Was the 2:1 ratio of length to diameter a general rule for varying pipe diameters and material combinations?' Mr Robson replied that in his experience the 2:1 ratio of length to diameter could safely be taken as a general rule for varying pipe diameters and material combinations. However, it was still important to insulate the flange joints between pipework constructed of different materials even with the internal cathodic surfaces coated to a distance of two diameters from the joint. He added that he knew of some experimental work in another organisation on the throw of galvanic current down small bore tubulars. He understood that this had arrived at a ratio of 3:1. Nevertheless, he was of the opinion that by using insulated joints a 2:1 ratio would be quite adequate even allowing for some deterioration of the insulation.

Dr D. Kirkwood (Robert Gordon's Institute of Technology) asked Mr Duncan (Paper 2.5, Cost implication of offshore platform structural protection by coatings) whether in view of the fact that paint failures arise from poor surface preparation and/or the prevailing atmospheric conditions his comment that an extra few pence in paint cost was really significant. The slight quality difference between paints was surely relatively less important than these other parameters. Mr Duncan, in reply, agreed that poor surface preparation was responsible for most coating failures. However, his point was that, by leaving the choice of coating to the applicator (from an approved list of coating manufacturers) it was inevitable that the overriding criterion would be the price per litre and the cheapest material was not likely to be the best. It might be better for the owner of the platform to select the material most suited to his requirements and issue it to the contractor free of charge.

The papers on stainless steels and other corrosion resistant alloys provided significant discussion. In commenting on Paper 2.6 (Cost effective use of stainless steel in topside module cladding and structural applications) Dr Gooch agreed with the author on the ready weldability of austenitic stainless steel and with the remark that weld decay should be regarded as a thing of the past. He asked for further information on the grades of stainless steel that had been used offshore for modules and on the history of these in terms of pitting resistance, etc. Dr Hooper said that Type 304 was the first steel to be used for module cladding offshore but current practice is to use Type 316 on all recent fabrications. So far there is no evidence of pitting – only rust staining is present. Mr Glover asked for the cost differential between the 304 and 316 grades. Dr Hooper indicated that Type 316 was about 35 per cent more expensive than Type 304 although the difference varied according to the product form and the molybdenum and nickel alloy surcharges which can fluctuate wildly.

In discussing Mr Glover's paper (2.7: Corrosion resistant materials for sea-water systems) Dr Hooper stated that because of the greater tidal lift and wave height in the North Sea compared with the American test site quoted the splash zone conditions would be very different and he would therefore ask for a definition of the size of the North Sea splash zone and a comment on the difference in the corrosion rates of steel. On the first of these points Mr G. Cammack, through Dr D. Fairhurst (BP Pet. Dev., Dyce) stated that their (BP's) definition of the splash zone for the northern North Sea is taken as 0-30 feet above Lowest Astronomical Tide (LAT). On the second point raised by Dr Hooper, Mr Glover said that higher corrosion rates are observed in the North Sea, compared with the test site in North Carolina, because of the higher dissolved oxygen levels in the colder sea-water and the erosive effects on any protective oxide films, which attempt to build up, by the violent wave actions encountered in the splash zone. Dr Fairhurst added that on the BP Forties platform jackets measurements of pit depths had provided an estimate of a minimum corrosion rate of around 0.5 mm per year in the splash zone. In a further comment Dr Hooper stated that stainless steel does not corrode quickly in sea-water if it is cathodically protected by e.g., welding to carbon steel. He asked Mr Glover whether, because of this effect, stainless steel could be used for splash zone protection. Mr Glover said that stainless steel when in direct contact with a less noble material such as carbon steel is cathodically protected in sea-water and could be considered for splash zone protection. Long term tests in the USA have shown that little pitting attack was observed on Type 304 stainless sheathing in the splash zone. Austenitic stainless steels are not recommended for riser pipe sheathing as there is certainly a strong possibility of stress corrosion cracking. Mr Glover was also asked by Dr Fairhurst to elaborate on Fig. 2.7,1 regarding the stainless steel bolts which had suffered severe corrosion with respect to the actual service conditions of their exposure. Mr Glover replied that both bolts had been sheathed in Nylon and had been in service for 15 months. He noted, however, that the conditon of the bolts was excellent outside the sheathed area.

The authors of Paper 2.8 (Application of novel anti-fouling coatings on offshore structures) were questioned on the abrasion resistance of the coating by Mr Bray – 'Were its properties impaired by abrasion of the surface?' They replied that the abrasion resistance of the material compares favourably with resinous coatings such as epoxies and urethanes. An exposure trial with two sets of panels, one of which was abraded, suggested a slight decrease in the materials' performance. However, practical exposure trials have indicated that mechanical damage to the surface of the coating does not impair its anti-fouling characteristics. Mr Glover asked for the cost of the paint system per m^2 and how this compared with conventional paint systems. The authors said that they could not give a precise answer to the question since the material is still, to a degree, under development. They expressed the view that it would be wrong to make comparisons with conventional anti-fouling paints as it was felt that this novel material would offer a much greater life expectancy. It was hoped that the cost data would be available later at a more appropriate stage but it was the authors' opinion that the cost would not be prohibitive.

REFERENCE

[3.1,1] Corpe, W. A. Microfouling: the role of primary film forming marine bacteria. In *Proc. 3rd Int. Cong. on Marine Corrosion and Fouling*. Edited by DePalma, J. R. and Iverson, W. P. National Bureau of Standards, Gaitersberg, Maryland, USA, 1972, 598–609.

Index